#HUMANRIGHTS

Stanford Studies in Human Rights

#HumanRights

The Technologies and Politics of
Justice Claims in Practice

Ronald Niezen

Stanford University Press
Stanford, California

STANFORD UNIVERSITY PRESS
Stanford, California

Printed in the United States of America on acid-free, archival-quality paper

Library of Congress Cataloging-in-Publication Data
Names: Niezen, Ronald, author.
Title: #HumanRights : the technologies and politics of justice claims in
    practice / Ronald Niezen.
Other titles: Stanford studies in human rights.
Description: Stanford, California : Stanford University Press, 2020. | Series:
    Stanford studies in human rights | Includes bibliographical references
    and index.
Identifiers: LCCN 2020002346 (print) | LCCN 2020002347 (ebook) |
    ISBN 9781503608894 (cloth) | ISBN 9781503612631 (paperback) |
    ISBN 9781503612648 (ebook)
Subjects: LCSH: Human rights advocacy—Technological innovations.
Classification: LCC JC571 .N527 2020 (print) | LCC JC571 (ebook) |
    DDC 323—dc23
LC record available at https://lccn.loc.gov/2020002346
LC ebook record available at https://lccn.loc.gov/2020002347

Typeset by Kevin Barrett Kane in 10/14 Minion Pro
Cover design by Christian Funfhausen

Contents

Figures

Foreword

Ronald Niezen's *#HumanRights* is both a deeply unsettling and profoundly urgent study of the very outer boundaries where justice making encounters the limits of technological possibility, agency, and even human sociality itself. In making the radical choice to focus his inquiry around what he describes as the "charisma of technology's power," Niezen's research proposes a wholly different approach to understanding many of the most pressing questions for the fields of human rights, science and technology studies, justice studies, and critical anthropology, including the origins of human rights in the twentieth century, the capabilities of new technologies to forge what Donna Haraway called a "cyborg politics" of machine-directed action, and the challenges for ethnography in a world in which representation, self-representation, and mediated curation have taken on entirely new forms.

The empirical foundations of *#HumanRights* are as heterodox and innovative as its theoretical and political implications. Ranging across a thrilling landscape of technological and communicative materialities, Niezen follows the evidence wherever it leads him: on walls in public spaces where graffiti appears from Paris to Cairo; through the monuments to collective memory from Namibia to Germany; on homepages of shadowy organizations that might or might not be working for progressive change; and even into the nanospaces where computer codes, processor capacity, and internet connectivity are replacing courts, human rights conventions, and democratic politics as the media and mechanisms on which claims for justice increasingly depend.

In this sense, the landmark methodological iconoclasm that forms the basis of *#HumanRights* should leave no doubt that real insight into the possible futures of human rights demands that we be willing to untether ourselves as scholars, policy makers, and activists from existing and often all-too-familiar verities. As Niezen shows us, we too must be willing to take risks in order to imagine the consequences for human rights of the technologically infused political, moral, and economic worlds that are fast upon us. Yet as *#HumanRights* also demonstrates, it is not always obvious which is the right cat to be belled, what is the real threat to human rights against which our attention should be collectively concentrated.

And this is yet another startling conclusion of Niezen's study: at a time in which the very meaning of truth is being undermined by cynical political actors at the head of ideologically unaccountable movements, the way forward is fraught with uncertainties. The return—at some distant point in the future—to a sociolegal context in which legal truth about human rights crimes and culpability is straightforwardly revealed through the traditional rules of courtroom evidence hardly seems likely. But as Niezen's research also shows, the technological ability to prove that brutally executed bodies lay on a geospatially determined stretch of sand on a known date at a fixed hour doesn't seem to be bending the arc of history any closer to justice either.

As Niezen argues, this is yet another paradox that increasingly shapes justice seeking in the contemporary world: although we now have the ability to develop thick dossiers of computer data, photographs, and videos that can admit of little challenge about what happened to which people on what day, this information does not, and *cannot*, lead inevitably to what might be called "algorithmic justice." As *#HumanRights* teaches us, claims for justice—including those grounded in human rights—don't actually become claims for justice until they are interwoven with the kinds of historical, cultural, and ideological meanings that can never be captured, let alone produced, by the instruments of the unfolding technological revolution.

And in exploring, with nuance and tremendous intellectual force, paradoxes like this, *#HumanRights* becomes something even greater than an incomparable reflection on what Niezen describes as the "technologies and politics of justice claims in practice." Instead, in examining with anthropological subtlety the permutations of the "dialectics of despair"—the ways in which our relationship to technology swings between the extremes of utopianism and dystopianism—Niezen's study is nothing less than a critical ethnographic treatise on the contemporary condition itself in all of its ethical aporia, concessions to (and dependence on) power, and desire for enclosure.

Mark Goodale
Series Editor
Stanford Studies in Human Rights

Preface

Whatever reality there might be to the hopes and fears circulating about the new information age, there is no denying the growing presence of new technologies in daily experience. Let me give a couple of personal examples. As I recently boarded a plane from Washington, DC, to Paris, I was surprised to find that my passport was unnecessary—cameras connected to a facial recognition application confirmed my identity and, one by one, that of every other passenger on the flight. (I couldn't repress troubling thoughts about how this technology might otherwise be used.) Not long after my arrival in Paris, I stopped by in person to make a reservation in a restaurant near where I was staying. I gave the woman at the desk my first name, and when she entered it into the touchscreen, I could see the rest of the information about me fill in automatically, including my home address and telephone number in Canada. "Ah, vous voilà!" she said, and I left with both a confirmed reservation and a puzzle as to how my information had found its way into the restaurant's database.

I'm sure that many readers of this book will have had their own encounters with such points of technological acceleration and dissonance. One of the odd things about these kinds of experiences is their ordinariness, their matter-of-factness. There is none of the fanfare or heroism of the kind described by Stefan Zweig in his account of the laying of the transatlantic telegraph cable that first connected the United States and Europe with instant messaging in 1858.[1] Innovations are simply finding (or obtruding) their way into daily life, with the struggles and heroism that lie behind them taking place hidden from view, in boardrooms and the minds of engineers and programmers.

In this book I am reappraising human rights through the lens of such experiences, or to be more precise, through consideration of the combined technological and political revolutions that are producing them and bringing them into the realms of both ordinary life and extraordinary struggles for justice. Everyday encounters with new information technology involve experience with the very tools being used by some states and corporations, alone and in collaboration, to invade the privacy of individuals and organizations, often with the intention of muzzling dissent.[2] This invasion of privacy includes the rise of social media as once-unimaginable sources of corporate profit and as platforms of misinformation and incivility that have become instruments of power.

Bleak as these developments might be, I did not find in the research for this book an absence of creativity and will toward freedom by those making human rights claims. There are new methods and powers to be found in technologies applied toward collective organization, public mobilization, and truth seeking. Taken together, innovations in information technology have produced a multidimensional digital arms race, most apparently one in which ambitious states are in competition with one another for supremacy over hacking capabilities and social media manipulation in struggles for power, but also in which activists and advocates are asserting basic claims and creating conditions of public persuasion, accountability, and respect for human dignity. In the rapid development of new information technologies and their application, it has become possible to see the contours of a new era of human rights, evident in both the means of oppression and of dissent.

In this book, I have assembled these developments in the form of an ambitious goal: to present a picture of a particular moment in time in the relationship between a rapidly changing media landscape and the processes and means by which people are attempting to bring the powerful to account. The only real point of clarity in this moment is its uncertainty, marked by areas of contest that remain undecided: the growing reach of "surveillance capitalism" that has not reached an end point;[3] the uncertain outcome of the post-truth crisis, as channels of information are readily captured and subjected to strategic distortion; and the efforts of the "tech left" and countless owners of smartphones to create new avenues to truth and accountability. A focus on human rights activism provides a point of departure to explore these contests and crises in a panoramic way.

Another part of what I have to say, particularly in chapter 2, is historical. There appears to be a connection in time and place between the so-called human rights revolution and the dramatic innovations that have taken place in information technologies. An argument can be made that these technologies have facilitated and shaped human rights and social justice advocacy in key ways, which largely account for the transformations that historians are pointing to as central to our understanding of where the international regime of human rights comes from and where it is going. I offer just such an argument here, through an ethnographically informed study of the contemporary.

My introduction to the work of Bellingcat, a citizen-journalist "collective" based in Leicester, UK, represents the furthest distance from my personal starting point in these explorations. In my participation in a workshop in Amsterdam and conversations with the founder, Eliot Higgins, the open source investigators conducting the workshop, and other contributors to the loose organization they refer to as "the Collective," I gained insight into and greater curiosity toward this emerging dimension of monitoring and forensics efforts oriented toward (among other things) encouraging state compliance with the standards of international law (as described more fully in chapter 3).

Human rights would diminish and disappear without the energy provided by its claimants, and it is on their perspectives that I focus, particularly in chapters 5 and 6. In taking this approach, I rely on case study–based illustrations of activists' navigation of their particular media ecologies, bearing in mind the anthropological injunction to the effect that it is important to begin from the ground up, to learn from social movements in action as they make use of various ITs and "traditional" media to organize, communicate, and mobilize, rather than to begin with the technologies and argue about their possible revolutionary effects. Consistent with this point, I have made an effort to hammer some longer-term research into new shape and significance, making it possible to offer two transnational case studies based on historical and ethnographic methods: that of the Tuaregs of northern Mali and their diaspora in Europe and that of the Ovaherero and Nama in a transnational campaign for genocide recognition in Namibia and Germany. Here my goal is to shift perspective toward those who are asserting justice claims in complex conditions, following the premise that justice-oriented technologies cannot be properly understood in abstract isolation, but can be better situated in the actual conditions of their development and use—that is to say, in the untidy mess of state power gone awry, competing political aspirations, memory and memorialization of mass atrocity, and struggles over public persuasion and sympathy. Focusing too narrowly on the tools of advocacy makes them stand out unnaturally, producing the same distortions as any other undue media attention to things that resonate with publics.

The inclusion of this kind of case material has made this book challenging on several levels. I am pulling together and recontextualizing some of my previous work (albeit sparingly, not wanting to repeat myself) and adding the results of new research, all while taking on a much wider topic: the way that new ITs are reconfiguring human rights advocacy and, in return, the way that this advocacy serves as a focused way to understand some of the social consequences of these technologies. My intention is to provide a picture in time of the new ecosystem of human rights activism without giving up on the ideals of closer insight and accidental discovery in context. I have hopefully been able to draw on just enough case material in this book to give a sense of the variety, creativity, and, at times, desperation of those who are using new media to convey their experiences and represent the conditions of their lives in efforts to raise public awareness of their claims and causes.

Expressing thanks to all those who contributed in some way to the production of this book is made difficult by the fact that my obligations extend back through the decades, so let me begin more recently and see where this takes me. With the freedom that came from a year as a visiting professor at Harvard's Weatherhead Center for International Affairs in 2018–2019, I was given exposure to new and emerging research at Harvard, including at the Weatherhead Center, the Department of Anthropology, and the Berkman Klein Center for Internet & Society at Harvard Law School. I am grateful to John and Jean Comaroff and George Meiu for giving me a place (actually, something very nearly like an intellectual home) in the workshop series at Harvard's Center for African Studies, particularly for the opportunity to present and receive outstanding feedback on a draft of my paper on Ovaherero and Nama rights claims in Germany, included here as chapter 6. My explorations during this leave year also took me to events hosted by the MIT Media Lab and MIT's Comparative Media Studies / Writing program.

Looking back a little further, I am indebted to the Max Planck Institute for Social Anthropology in Halle(Salle), Germany, for hosting me often and offering that rarest of things: quiet in the context of rich library resources and scholarly ferment. In terms of funding, I had the great advantage of flexible support from two chair programs: a Canada Research Chair in the Anthropology of Law and the Katharine A. Pearson Chair in Civil Society and Public Policy. The organizations that hosted me as a researcher include Bellingcat, the Organisation de la Diaspora Touaregue en Europe, the Ovaherero Genocide Foundation, Berlin Postkolonial, SITU Research, and MIT Students Against War. My understanding of the work of these organizations would be incomplete without their members who shared their experiences and insights with me. I benefited from presenting my work-in-progress at Harvard University's Department of Anthropology and the Weatherhead Center for International Affairs, Sciences Po Lille, the University of Baltimore, Johns Hopkins University, and the University of Essex. Two chapters of this book are much-revised versions of work published elsewhere; I owe thanks to Cambridge University Press, which published my chapter in Sandra Brunnegger's edited volume *Everyday Justice: Law, Ethnography, and Context* (appearing here in chapter 1), and to the *International Journal of Heritage Studies*, which generously allowed my material to appear here (in chapter 6). I am grateful for the support and involvement of the people at Stanford University Press, above all my editor, Michelle Lipinski, and the anonymous reviewers of my proposal and manuscript for their important suggestions.

This book derived much of its original inspiration from a 2016 panel at the annual conference of the American Anthropological Association, put together by Rita Kesselring, in which participants were asked to imagine the future of the anthropology of law. I am grateful to Maria Sapignoli for thinking through this problem with me and following up with important conversations (often supported with references) about information technologies in law and global governance. Rachel Thompson went beyond her work as a research assistant by introducing me to people and panels at MIT that I otherwise would have overlooked and commenting on a draft of the manuscript. I also received feedback on chapters of the draft manuscript from John Comaroff (chapter 1), Reinier Torenbeek (chapter 2), and Enrique Piracés (chapter 3). From here, the list of individuals who in one way or another contributed to this project and whom I haven't already mentioned includes Mariam Aboubakrine, Payam Akhavan, Jennifer Allison, Abdoulahi Attayoub, Nick Barber, Chris Bavitz, Adelle Blackett, Josefina Buschmann Mardones, Emily Chow Bluck, Matthew Canfield, Gabriella Coleman, Alonso Espinosa-Domínguez, Henk van Ess, Lindsay Freeman, Mark Goodale, John Hall, Eliot Higgins, Bob Hitchcock, Ian Kalman, Stuart Kirsch, Arthur Kleinman, Alexa Koenig, Michele Lamont, Gaetano Mangiameli, Félim McMahon, Chukwubuikem Nnebe, Pierre Peraldi-Mittelette, René Provost, Tobias Rees, Annelise Riles, Sally Merry, Sally Falk Moore, Colin Samson, Niels Schia, Nicole Rigillo, Brad Samuels, Christiaan Triebert, Bertram Turner, Noah Weisbord, Richard Wilson, and Olaf Zenker. To those who find their name missing from this list, it means you contributed in some way without my acknowledgment, which puts me even more in your debt.

Finally, saving the best for last, Sarah Federman was with me every step of the way through this project, offering inspiration, insight, patience, photographs, and boundless, infectious enthusiasm for life.

Abbreviations

ACTUP	AIDS Coalition to Unleash Power
AfD	Alternative für Deutschland (Alternative for Germany party)
AI	artificial intelligence
AQMI	Al Qaida au Maghreb islamique (Al Qaida of the Islamic Maghreb)
ARPANET	Advanced Research Projects Agency Network
ASSÉ	Association pour une solidarité syndicale étudiante (Association for student union solidarity)
BFDT	Bündnis für Demokratie und Toleranz gegen Extremismus und Gewalt (Alliance for democracy and tolerance against extremism and violence, Germany)
BBC	British Broadcasting Corporation
CBS	Columbia Broadcasting System
CCTV	closed-circuit television
CEO	chief executive officer
CMA	Coordination des Mouvements de l'Azawad (coordination of Azawad movements)
CONPAZ	Coalition of Non-Governmental Organizations for Peace
CRPD	Convention on the Rights of Persons with Disabilities
DIVA-TV	Damned Interfering Video Activist Television
DoS	denial of service
DRS	Département du renseignement et de la sécurité (Department of Information and Security, Algeria)

DVD	digital versatile disc
ECHR	European Court of Human Rights
EZLN	Ejército zapatista de liberación nacional (Zapatista National Liberation Army)
GDR	German Democratic Republic (former East Germany)
GLAN	Global Legal Action Network
GPS	Global Positioning System
GRU	Main Intelligence Directorate (Russian Federation)
IACHR	Inter-American Court of Human Rights
ICC	International Criminal Court
ICCPR	International Covenant on Civil and Political Rights
ICESCR	International Covenant on Economic, Social and Cultural Rights
ICJ	International Court of Justice
ICT	information and communication technology
IED	improvised explosive device
IETF	Internet Engineering Task Force
ILO	International Labour Organization
IoT	Internet of Things
IPACC	Indigenous Peoples of Africa Coordinating Committee
IT	information technology
ITU	International Telecommunications Union
MAR	Minority at Risk
MDG	Millennium Development Goals
MH17	Malaysian Airlines Flight 17
MIME	multipurpose internet mail extensions
MIT	Massachusetts Institute of Technology
MNA	Mouvement national de l'Azawad
MNLA	Mouvement national de libération d'Azawad
NAFTA	North American Free Trade Agreement
NATO	North Atlantic Treaty Organization
NGO	nongovernmental organization
NSF	National Science Foundation

NSFNET	National Science Foundation Network
ODTE	Organisation de la diaspora touarègue en Europe (organization of the Tuareg diaspora in Europe)
OGF	Ovaherero Genocide Foundation
OSINT	Open Source Intelligence
PETA	People for the Ethical Treatment of Animals
RAT	remote access trojan
RSF	Reporters sans frontières (Reporters Without Borders)
RT	*Russia Today*
SCCS	Swarthmore College Computer Society
SNCF	Société nationale des chemins de fer (French National Railway Company)
SOE	Special Operations Executive
SUV	sport-utility vehicle
SWAPO	South West Africa People's Organization
TV	television
UN	United Nations
UNDESA	United Nations Department of Economic and Social Affairs
UNDRIP	United Nations Declaration on the Rights of Indigenous Peoples
UNHCR	United Nations High Commissioner for Refugees
UNICEF	United Nations Children's Fund
URL	universal resource locator (i.e., web address)
VPN	virtual private network

Utopia and Despair

Those who are most enamored of social order must, to remain strong, tolerate, even promote dissidence and opposition; because, to remain enthusiastic and based on common belief [the political order] needs a constant influx of new discoveries and initiatives, that prod and rouse it with the sharp end of their strangeness.

Gabriel Tarde, *La logique sociale*, 1895

The Dialectics of Despair

Inventions often have a charisma that exceeds their utility. They invite positive values, welcome new possibilities for human betterment, feed the fertile imaginations of inventors and revolutionaries, and provide tools for dissent and different thinking. Like all charismatic figures, the power of new technologies sometimes has a fascinating element of unpredictability, of possibilities for spilling over whatever boundaries of control we might construct or imagine. True believers will bracket warnings out, refusing to see or accept the possibility of danger in a force with such potential for good, and hold on steadfastly to hope as a complete and permanent answer to the perils that stand directly before them.

This is especially clear with the historic jolt brought about by the emergence of the interrelated technologies of artificial intelligence (AI), smartphones, and social media. Despite their potential to bring about dramatic change and imaginings of a new world, and despite their occasional personification, the charisma of new information technologies (ITs) is not the same as that of the prophets and cult leaders of earlier times. The popularity of technical innovation either takes off or it doesn't—no need for the sword, the whip, and the auto-da-fé to keep a loyal following. The charisma of technology's power lies almost exclusively in its ability to maintain recognition by "proving itself" with remarkable, seemingly supernatural feats, on which its devoted followers base their unquestioned loyalty.[1]

An example of this kind of blind attraction to innovation comes from Miguel Luengo Oroz, chief data scientist for Global Pulse, a joint United Nations–Google initiative that uses big data, artificial intelligence, and other ITs to advance the goals of humanitarian intervention and policy implementation. Oroz is inspired in his work above all by a vision of humanity unified by the surveillance capacities of emerging technologies: "Before 2030," he proclaims, "technology should allow us to know everything from everyone to ensure no one is left behind. For example, there will be nanosatellites imaging every corner of the earth allowing us to generate almost immediate insights into humanitarian crises."[2]

Another example comes from the UN's Department of Economic and Social Affairs, which, in a newsletter sent out regularly via email to subscribers, offers an unabashedly Panglossian view of the emergence of what it calls "frontier technologies":

> Frontier technologies are innovative and often grow fast, with the potential to transform societies, economies and the environment. In recent years, we have seen examples of this in the form of artificial intelligence and machine learning, renewable energy technologies, energy storage technologies, electric and autonomous vehicles and drones, genetic engineering, as well as cryptocurrencies and blockchains. These frontier technologies can help eradicate hunger and epidemics, increase life expectancy, reduce carbon emissions, automate manual and repetitive tasks, create decent jobs, improve quality of life and facilitate complex decision-making processes. In other words, these technologies can make sustainable development a reality, improving people's lives, promoting prosperity and protecting the planet.[3]

From this, it seems, one would need nothing more to bring about a perfect world than to harness the inevitable force of technical invention and apply it to projects for the social good.

For many who read these lines, the powers of frontier technologies do not offer unreserved comfort, never mind visions of utopia. The idea of nanosatellites probing every corner of the earth and transmitting information on our every movement invokes in our imaginations the more sinister uses of technology. Chillingly, we are not told what person or agency will be receiving this information about our every movement. All that seems missing is some mechanism to probe and transmit our thoughts—perhaps by 2050.

Technological optimism, however, cannot be assumed as a natural accompaniment of innovation. Writing on the history of technology in the immediate aftermath of World War II, a time when the power of invention had been on full display and invited critical reflection, Sigfried Giedion noted a climate of skepticism toward the tools of modernity: "Now," he writes, "it may well be that there are no people left, however remote, who have not lost their faith in progress."[4] The trend that Giedion observed is still part of the way that many respond to innovation in the technology sector: almost reflexively understanding that under the control of centralized power, ITs serve as tools for abuse by those seeking or securing centralized power. Advances in the uses of big data mean that liberties can be trampled underfoot, minds corrupted, faiths destroyed, new methods of violence put into action. Ultimately, by small degrees, according to the most jaded view, revolutionary innovations in technology bring in their wake the mobilizations of armed men and smoke rising from villages.

Sometimes, however, individuals are torn between these possibilities and wage an inner struggle for the primacy of one attitude over the other. While recognizing the power of a technology to alter human lives, they just can't be sure if the menace they see is real or if lilies are about to bloom in the desert. This is the struggle that, for want of an existing term, I call the dialectics of despair. It is a theme relating to new technologies that gets evoked constantly, and it came up so often in the writing of this book that I usually had to leave it on the page without comment. It is one of those ideas that are in the ether of the era.

Alternation or indecision between the ideas of an imminent media-induced utopia or a condition of dystopic human enslavement is now shared by contemporary analysts of the emerging social conditions being shaped by new ITs, except that in mainstream media at least, the dismal view seems to have gotten the upper hand. Many analysts went through a period of euphoria at some point in the early 2000s as innovations leaped ahead, creating new forms of engagement, connection, and organized dissent. Many technology experts and activists were thrilled at encountering the potential of new ITs to create networks and social movements with unprecedented reach and power. They seemed to have picked up where Marshall McLuhan left off in *The Gutenberg Galaxy*, with his almost breathless anticipation of the age of computers, which "promises by technology a Pentecostal condition of universal understanding and unity" in which languages are bypassed "in favor of a general cosmic consciousness."[5] We can safely assume that those writing in the full bloom of the internet era were not optimistic, as McLuhan was, in the way of anticipating an imminent global hive mind "that extends our senses and nerves in a global embrace"[6] but more with a sense of a coming era of libertarian freedom, with messaging, networking, and open source platforms celebrated for their capacity to infuse capitalism with creative vitality on a utopian scale or to enable effective resistance against corporate abuses and state violence while ushering in new collective values.

Then came the fall. The post-9/11 architecture of state intelligence gathering and the descent of the Arab Spring into hyperrepressive governments and the civil wars of Syria and Libya were major shocks experienced by many tech savvy activists. Encroachments on privacy, the virulence of anonymous messaging, and the development of new, more powerful practices of surveillance—these are continuing problem areas highlighted by a more cynical take on technologies, with important implications for those concerned with the future of civic freedoms and human rights. A main current of analysis and media reporting of technology takes the form of a morality tale of oppression/opposition in which unseen and unknown powers are aligned against digitalized activists, prepared to leap into collective action in campaigns of "hactivism," whistle-blowing, and "netizen action."[7] As a result of this framing, when it comes to analyzing the technologies that have implications for human rights, opinions tend to be committed in one direction or another in a with-us-or-against-us sort of way. Even if scholars try to be nuanced or at least to give opposing viewpoints their due, people tend to lean toward visions of either technologically reinforced structure or empowered agency, either fatalism or hope.

What appears to be inconsistency in these views, however, is actually an outcome of inherent and unresolved contradictions in the dynamics between innovation and power. We are living in a time in which the technologies bringing about sweeping change—with new ones in "alpha" and "beta" stages of being tested, waiting in the wings and soon to be introduced—have not fully coalesced into clear patterns. Some state and corporate actors are subjecting those engaged in protest and dissent to the uses of technology in surveillance, censorship, and the often-violent narratives of strategic falsehoods. At the same time, dissidents are putting a spotlight on these invasive structures of state and corporate power through new forms of protest, policy, and counterinvention—the "digital civil society" that has become necessary for the new technologically enabled powers to be checked and kept within bounds.[8] Or, to express this tension in another way, technologies of legal advocacy are facilitating new forms of transnational organization and public outreach while, at the same time, these very technologies have begun to transcend human will and agency, creating vacuums of transparency and opportunities for domination, surveillance, and stifling of dissent. A digital arms race between states and dissidents, mediated by big tech corporations that are playing both sides, has yet to be decided. We have every reason to hopeful and every reason to be afraid.

The new and emerging conditions I have just described provoke a key question: What happens if, instead of considering human rights only in juridical terms, we were to also look at them from the perspective of the *technologies of human control and persuasion*? This brings me to the central thread that connects the various chapters of this book: Within the past decade (or just a bit longer), a dramatic shift has taken place in the technological environment in which human rights are situated. Computers with unprecedented speed and capacity, together with innovations in artificial intelligence and machine learning, are at the heart of a technological revolution that is shifting the foundations of knowledge and public influence. The powers of repression and dissidence have each been amplified, ushering in a new era of surveillance capitalism,[9] digital state power, and, struggling to keep pace, war crimes investigations and human rights advocacy.

In this book, I situate human rights within a technologically thick setting of contest and contradiction. It represents an effort to put together some of the main attitudes to transformation in human life and possibility, each with its own discourses of utopia and despair. Human rights provide a lens that magnifies and brings focus to these attitudes toward the nexus of new technologies and old freedoms. They offer a clearer sense of what is at stake and why passivity is not an option for those committed to preventing a dystopian future, of either the Orwellian (the extended reach of state power) or Huxleyan (the cultivation of a comfortably numb, passive compliance) kind.

While the first inclination in response to the developments I have just outlined is to focus on the "shiny new things" of technological applications—including the concerns that they elicit—I propose to also take an approach that is grounded in personal encounter with justice claimants and their messages. Starting with the justice claims of the marginalized provides a more realistic understanding of how ITs are being used, in conjunction with a wide array of other technologies, in the interests of countering the narratives of states and bringing the claims of the marginalized into public recognition. Considering what justice activists actually do will often reveal a wide spectrum of media uses, far more diverse than the stereotype of cyberactivists equipped with mind-boggling computer systems and skills (though I discuss this kind of actor too). Ultimately, some of the oldest media used to sway opinion—including graffiti, placards, monuments, and printed leaflets—are being used alongside social media and mainstream journalistic media coverage (still the ultimate prize) in campaigns that promote such causes as peaceful solutions to the main human rights challenges, including armed conflict targeting civilians, forced displacement, genocide, and inhumane treatment of refugees.

Technology versus Values

Unless you've been living under a rock, you don't have to be reminded that new information and communication technologies are influencing life on the planet in unprecedented ways. According to 2017 Gallup World Poll data, an estimated 93 percent of adults in high-income economies owned a mobile phone, while 73 percent did in developing economies.[10] (To put this in perspective, 70 percent of people worldwide in 2015 were living on less than $10 per day, which, astonishingly, did not prevent most of them from acquiring a phone.)[11] By 2021, the number of mobile phone users is expected to exceed seven billion, with nearly half of their devices being smartphones.[12] The poorest of the poor may or may not have a cell phone, but they all know someone who does, making the connectivity of the world very nearly complete.

The International Telecommunications Union (ITU) of the United Nations still talks about a "digital divide" that separates the more and less connected countries and regions and that leaves rural areas less well connected than cities. But the overall trend has been toward rapidly increasing internet access, with more than half (51 percent) of the world's population connected as of June 2017.[13] (This compares with roughly 17 percent in 2005, marking an exponential jump in the number of people online.) Worldwide, the speed of connectivity is getting much faster too, with fixed broadband services (those having speeds of 256 kilobits per second [kbits/s] and above) having increased by 183 percent between 2007 and 2017.[14] Chances are these figures and the technologies they represent will seem laughable decades from now, especially with the near-future construction of a 5G infrastructure and new potentials for quantum computing—which would exponentially increase computing speeds and capacity for data storage—now part of a race for development. But this still leaves us with a revolution of sorts to contend with, in which the world has connected with a rapidity that, in global historical terms, amounts to a nanosecond.

This explosion of mobile telephony and online access has been accompanied by a range of new uses of communication and information technologies in human rights work. The mere fact of the omnipresence of digital cameras, for example, has increased the possibilities of human rights reporting. Social media sites like Facebook, Twitter, Instagram, Weibo (in China), YouTube, and many smaller platforms have made it possible for those who witness or experience violence to post information about it online—within the limits of state-sanctioned censorship and repression, of course. Organizations such as Ushahidi, CrisisMappers, Witness, and The Whistle enhance the ability of ordinary people with extraordinary technologies to produce data about human rights violations whenever and wherever they occur. The open source research collective Bellingcat (discussed further in chapter 3) is leading the way in making use of online data to assemble meticulous and reproducible reports, often pointing to both mass atrocities and clumsy efforts to cover them up. More than ever before, acts of violence are being recorded by witnesses, posted online or forwarded to journalists and advocacy groups and collated into patterns of abuse that point to those responsible.

This is not the first time the world has shrunk in unprecedented ways, and I have drawn inspiration from earlier eras of invention and their interpreters. Gabriel Tarde, a French jurist and philosopher of the late nineteenth and early twentieth centuries, can be credited with the first extended reflections on the connection between technologies of communication and the formation of nationally oriented mass publics.[15] The era in which he lived most of his life and formed his ideas, the latter half of the nineteenth century, saw the rise of new journalism in Europe and North America, marked by appeals to mass audiences, which naturally enough went together with a rise in sensationalism, sentimentalism, and nationalism in the subject matters of newspaper and periodical articles.[16] Oddly, though, Tarde was largely by himself in his identification of the telegraph, railroads, and newspapers as centrally significant for the study of human interaction. As Norbert Elias put it, "Hardly anyone spoke clearly about the rapidly increasing integration of humanity. . . . Thus, the shortening of distances, the increasing integration, happened, as it were, in secret. It did not obtrude itself on human experience as a global process of integration."[17]

As the great innovator of communication theory in the late nineteenth century, Tarde had to decide what to make of it all, especially concerning the ways that technology influenced social interaction. On the one hand, the increased range of human interaction meant that innovations reached wider publics, with the possibility for greater impacts on a larger number of individuals. And that process of expansion, in turn, meant that countless other institutions, practices, and beliefs would be abandoned, inevitably reducing the range and repertoire of human variety. "The work of journalism," he observed, "has been to nationalize more and more, and even to internationalize, the public mind."[18] Oddly to our contemporary countermodernist sensibilities, Tarde was not nostalgic about this turn of events. This was simply in the nature of things. His emphasis on publics, as constellations of individuals brought into relationships through networks of communication, made it much easier to imagine a common humanity unified by the growing reach of ideas of mass appeal and the institutional adaptations (or revolutions) that follow from them.

What about the centralized tendency to control ideas? Were there not new possibilities for ideological authoritarianism built into the new technologies of communication? As is suggested by the epigraph to this chapter, Tarde identified social and political progress with dissent, with the provocations of ideas that prod and rouse "with the sharp edge of their strangeness." At the same time, he seemed blithely unaware of the forces of propaganda lurking within the very media he wrote about, which were later to have such calamitous effects in the twentieth century. For him, reflecting and writing in a more innocent time, publics were consumers, sometimes astute, at other times misguided, that form around common worldviews. The commercial uses of media were one major source of his disapprobation: "The inducements of fairground charlatans extend no further than the reach of their voices; those of the charlatans of signs and advertisements seek out their dupes in an immense region."[19] He was also acutely aware of abuses of information in popular journals, which he once referred to as "these gigantic bellows that stir up passions from below."[20] He articulated a clear awareness of the credulity and docility of publics, whose imitations can be unconscious and involuntary: "We must deplore the inventive genius expended on clever lies, specious fables, all continually contradicted, continually revived, for the simple pleasure of serving each public the dishes it desires, of expressing what they think to be true, or what they wish to be true."[21] Reacting against such trends, he even went so far as to advocate a style of journalism that avoided political commentary altogether and presented its readership with only dry short articles and statistics.[22]

This idea of factual reporting as a solution to the failings of popular media was to come up again in critical analyses of public opinion. Walter Lippmann, writing in the immediate aftermath of World War I, for example, was more acutely aware than Tarde of the abuses of propaganda and the intellectual frailties and susceptibilities of public consumers of information. At this point in the early twentieth century, it was easier to see the artful refinements that generals and politicians applied to what he called the "manufacture of consent" (largely as a way to conceal and/or justify the massive casualties of the war) and the receptiveness to propaganda by the "anonymous multitude." For Lippmann, the way out, the path to "betterment," lay in "intelligence work," by which he meant expert knowledge based on the efforts of those who "go out to find the facts and to make their wisdom."[23] In this conclusion, he is not so very far from the contemporary advocacy of open source intelligence as a solution to state-sponsored manipulation and misinformation in social media (a topic of chapter 3).

There are strong currents in more recent literature to support a technology-centric approach to new forms of dissidence and change along the lines of Tarde's and Lippmann's musings on the sources and consequences of mass communication.[24] As Neil Postman puts it, "the weight assigned to any form of truth-telling is a function of the influence of media of communication."[25] Comparative anthropologists, risking the inevitable critique that follows the shift away from tightly focused ethnography toward wider concerns, have also taken up the question of the technology of communication as a driving force of human history. Jack Goody, for example, made a name for himself as a theorist of the consequences of alphabetic literacy and typography, considered both historically and with reference to contemporary societies. In a major body of work on the consequences of literacy, he made efforts to understand the influences of alphabetic writing on bureaucracy, dissent, and the growth of knowledge.[26] His work in this field constituted one of the more important inquiries into the ways that the technologies of literacy and typography influence human cognition and social order.[27]

This book takes on the question of the social consequences of information and communication technology with attention to the realm of human rights. I reconsider the history of human rights through the intersecting ontologies of communication and law in the global production and networking of justice-oriented social movements. This reconsidered history will not replace one form of determinism with another. I begin with the premise that new technologies, just like the rights to which they are applied, are anchored to values that become expressed through media of communication. When it comes to deciding how technologies might act to influence justice causes, we should be ever cognizant of the complex multiplicity of values in which rights are embedded and contested, and the often-creative consequences of their collision.

At the same time, the technologies associated with persuasion are every bit as important in amplifying (or, less often, shaping) opinion, and ultimately influencing the outcome of justice causes, as the information, arguments, and values they convey. The most visible starting point of the new era of justice lobbying can be found in the tools of communication and public outreach that have given dissidents a wider field of influence and concern. Artificial intelligence and machine learning are increasingly used in the service of legal process, including their application to human rights claims, monitoring, and forensics. And, by enabling unprecedented forms of global networking and collaboration, social media have become an integral part of justice claims, including through the principle of "sunlight as a disinfectant."

This principle, picked up (before it had a name) from the mainstream media reporting of the civil rights era, became central to the grassroots activism of the human rights movement in the 1980s and 1990s and continues in different livestreaming and web-scraping forms to this day. This observation brings me to a methodological point: One of the keys to better understanding the scope of technological revolutions is not to focus exclusively on the newness of things but to also be attentive to continuities and impediments to change. From this perspective, I hope to show that in the pursuit of justice, old forms of communication are dusted off and combined with the new in technological potpourris of activist outreach.

Digital Inequality and Activist Elites

I was once (in the spring of 1999, I think) travelling on a remote highway cut through the boreal forest of northern Manitoba, sandwiched in the back of a double cab Ford F-150 with a group of Band councilors from the First Nations reserve community of Cross Lake. They were on their way to a meeting with government officials in Winnipeg, and I was travelling with them as an observer, eventually to report the events of the meeting to the Band's lawyers. We settled into the boredom of a ten-hour drive, with little to see other than endless vistas of spruce trees stunted by Northern winters. Then, seemingly out of the blue, everyone in the vehicle—including the driver, with one hand on the wheel—simultaneously took out flip phones, punched in numbers, and began talking, each covering their exposed ear and competing to be heard. The truck suddenly sounded like a call center, which, in a manner of speaking, it was. What was going on? When the first person wound up his conversation and snapped his phone shut, I asked about the sudden flurry of calls. The reason was simple. A distance marker that they all knew and were waiting for provided a point of reference for connectivity, where the microwave-relay telephone system could be counted on to make a cell phone call that would get through and wouldn't drop.

Limited access to communication technologies, which was certainly prevalent in the late 1990s and continues to a somewhat lesser degree today, was a major obstacle to communication and networking in the northern village from which the Band officials came. The phenomena referred to as the "digital divide" or "digital inequality"—which include the uneven distribution and access to connectivity, communication tools, and digital media literacy—are still very much part of the media/activist landscape, but in ways that are changing rapidly. One of the current tropes of media analysis involves observations to the effect that only a few years ago, X people lived in isolation and poverty, but look at them now, all texting and connected, with everyone hunched over a smartphone. Ironically, the social dislocations that might follow from the spread of new technologies and networks everywhere in the world are offset by the fact that virtually everyone now has access to those technologies, or wants to. There are of course still those who value simple subsistence-based lifestyles far from contact with the wider world, but at the same time humans are everywhere remarkably social and there is little that separates people in terms of the values of access to technologies of connectivity, only the means to achieve it.

A second point that the connectivity of the Band officials raises for us has to do with the elite status of those who represent communities of human rights claimants. These men were all equipped with phones and the ability to travel and stay in hotels, while many of their constituents were not. It is impossible to fully understand the changing nature of human rights advocacy without taking into account the ways that leaders achieve recognition and act on their privileged circumstances through access to new ITs. This privilege allows them to connect with one sector of their constituencies—those who have the education, wherewithal, and technological means to maintain contact with them—while others become ever more marginal. Along with these technologies, a new stratum of computer literati is reshaping the status hierarchies, resistance strategies, and conceptions of collective self of many communities seeking recognition of and action on their rights.

The privileged status of human rights advocates is a variable that has shifted, mainly in ways that are connected to changes to the ecology of ITs. Those who advocate for the rural poor often have to contend with jealousies and misunderstandings that stem from their privileged access to technology and travel. Those who are now part of the small pool of software engineers working for big tech companies are an altogether different kind of elite, whose skills and privileged positions of employment give them an important and growing place in the emerging field of interconnections between technology and justice. To properly unpack the media ecology of the new era of human rights, we need to go in both directions—to the leadership of the marginalized and to the corporate/educated elites—to delve into the ideas and actions of both of these very different kinds of actors and the causes they represent.

A User's Guide to Human Rights

The main source of my experience with the nexus of media technologies and human rights comes from my work with marginalized peoples and communities, particularly the field of the human rights activism of indigenous peoples. As the years turned into decades (I started this work in the early 1990s), I was able to witness shifts that took place in the movement, including in the ways that delegates to international meetings made use of new technologies to communicate with their constituents, created and navigated global activist networks, and reached out to UN officials and journalists from major media outlets.

In this book, my attention to the indigenous peoples' movement is less sustained than it was in much of my previous work, though it is still there in the background or in shreds and patches in my observations on popular movements and social outreach. My central goal is to contextualize and capture a moment in the history of human rights and technology, and to do that I have had to cast a wider net. I have done this in several ways. Chapter 2 offers an overview of the seventy years of the human rights movement through the lens of technologies of activism. My subject matter ranges through various stages of the IT-rights connection, from the media outreach of Nazi hunters, through the Zapatista rebels and networked activism of the 1990s, to the thickly technologized world of social media activism and algorithmically mediated rights (and state control) that is emerging today. Here I take the theme of the emerging era of human rights activism and dwell a little longer on the new challenges and opportunities for public outreach in the contemporary media ecosystem. This is the point at which I have entered into the (to me heretofore) less familiar wilderness of tech savvy activists, tech experts, and the technologies they are using and developing. These expert-activists were of two kinds: the boots-on-the-ground activists who use social media, livestreaming, and social networking websites to amplify and make immediate the messages of their causes, and those who specialize in online forensics.

In chapter 3, I pay particular attention to this latter realm of tech expertise. This included an introduction to the forensic adventures of Bellingcat, centered on a one-week workshop in Amsterdam, preceded by exploration of its regular online newsletters and email exchange and followed by further online research, exchange, and interviews. Here, I learned, among other things, techniques for gathering material from obscure corners of the internet and collecting it into local databases where it can then be subjected to forensic analysis; but I also learned from my teachers and coparticipants something about their conceptions of justice and aspirations for applying new technologies toward its realization. Even the ideals of investigative rigor and transparency cannot entirely overlay the emotions and motivations behind activist efforts: indignation toward abuses of power, sympathy for its victims, and hopes for a world in which technology is a source of truth, counterempowerment, and a new global order.

The Publics of Public Outreach

International law, including human rights law, encourages global legal order by conventions and treaties that are entered into voluntarily by states, by monitoring and oversight done by specialized agencies, and by social pressure based on popular opinion.[28] It is this last ingredient—the most mysterious and amorphous and the least understood—on which I concentrate in this book. The ways that treaties are negotiated, hammered into shape, and entered into and the ways that international agencies set about their work of monitoring compliance are more or less readily understood, despite institutional obstacles of secrecy and invisibility. But getting ahold of the ways that publics might influence international law, with their amorphousness, ranges of opinion, variations of influence, and shifting alignments, is another thing altogether.

Most legal scholars recognize the significance of opinion and persuasion in the way that human rights work, including the pressures toward compliance that may or may not be effectively brought to bear on states and, occasionally but increasingly, corporations. But if this is so, then what or who are the publics acting to make human rights work? In addressing this question, we should be open to the possibility that the public opinion (or opinions—there is clearly a plurality) that gives effect to human rights might not correspond in a one-to-one way with the legal standards and official discourse of those rights. If "popular will" is central to human rights compliance, what are the ideas about universal justice expressed by that will? What forms does it take? How is it influenced by and what influence does it have on the law? There may be popular aspirations and ideals that contribute indirectly to formal conceptions of human rights through the opinions of inter-ested, activist publics that have not been accommodated in the formal standards and procedures of international law. Whatever goals global policy makers might envision as right and good for the world, people have their own ideas. And these might not have a comfortable fit with the way the UN is structured or the opin-ions its officials would prefer to hear. There is something to be gained by being attentive to the untidy, often contradictory, and incomplete ideas and efforts that, taken together, make up popular imaginings of human rights.

Awareness of the power of public opinion and the ways to capture it has become a significant part of the phenomenon sometimes known as the human rights revolution. It was not until the late twentieth century, when new media were used by a burgeoning number of human rights nongovernmental organizations to engage in global campaigns of networking and consciousness-raising, that we saw the full effect of organized protest movements seeking the attention and sympathies of public audiences—and the simultaneous efforts by those against whom they were protesting to silence them.

The Human Rights Committee, which monitors compliance with the Interna-tional Covenant on Civil and Political Rights, has always had as its first order of business in approaching a new claim the task of determining jurisdiction—that is to say, the question of whether the committee could hear the case after the exhaustion of domestic remedies. A claim is more often decided by judgment on the matter of whether the issue at hand had thoroughly wended its way through the states' court system than on the substance of the claim. The same also applies to cases heard by the European Court of Human Rights (ECHR) and the Inter-American Court of Human Rights (IACHR). Human rights pick up where state judiciaries fail. The formal procedures of human rights compliance understand-ably focus on separating human rights claims from their entanglements in state courts and, by implication, state conceptions of justice.

Popular conceptions of universal justice, however, follow no such process; and for this reason I try to avoid a narrow conception of human rights as defined by the normative commitments and processes of international human rights law.[29] Human rights compliance begins with public ideals outside courts and human rights venues, with the dynamics of exposure, outrage, and embarrassment of the accused, often followed by state repression and further outrage. Publics usually don't care if an injustice is a violation of human rights or criminal procedure or any other dimension or jurisdiction of law. The language of human rights in public conceptions of justice rarely stands alone. They become intertwined with notions of such things as violated indigenous treaty rights, rights of due process, or even environmental and animal rights. There is a transnational, sometimes global plebiscite that gives some justice claims the energy and resources needed to see a cause through its course in formal venues and to provoke political turmoil and possibly reform and remedy along the way. Public opinion, in which laws, aspirations, and jurisdictions are combined in complex webs of indignation, emotion, and misinformation, is the ultimate arbiter that gives human rights their legitimacy and powers of compliance.

Sometimes, in the midst of a struggle for rights, the concept of humanity fades into the background as the victims of rights abuse dwell on their particular circumstances of suffering and injustice. In the international meetings of indigenous people held at the UN headquarters in Geneva and New York, for example, it is easy as an observer to come away with the impression that the indigenous peoples' representatives have difficulty seeing beyond the specific circumstances of injustice that they face. Whenever the agenda opened up for speakers to address their human rights concerns at the Permanent Forum on Indigenous Issues, the lineup at the registration table would suddenly grow. Delegates were anxious to talk about how human rights affected *them*, how their state violated *their* peoples' rights. They would even try to slip their human rights grievances obliquely into discussions of other topics. They remained steadfastly focused on specific, often personally experienced injustices. Yet, on closer consideration, there was always in the background to specific grievances a sense of what a better world should look like, often in the form of how industrial extraction, displacement, and destruction of the natural world was impoverishing human life in general.

Given the significance of opinion in influencing the outcome of particular rights abuses, it is important to know how human rights are understood by those who are expressing opinions and exerting pressure toward compliance, even outside the ideas and edifices in Geneva and New York. From this perspective, it becomes possible to see that human rights are found not just in the instruments or treaties that give them formal expression, or in the work of the Human Rights Council with its regular meetings and special sessions about crises of the moment, or even in the activities and reports of the Universal Periodic Review, in which states offer regular assessments of the human rights records of other states. Human rights also have a life outside of these formal ideas, structures, and procedures. They are simply what humans think of as the rights that they share with all other humans.

Digital Ethnography

Human rights are divided between those who develop and administer these rights in agencies of global governance and those who are their claimants and defenders—between the "headquarters" and the "field." This is troublesome for people like me who have a professional identity as an anthropologist to keep up, who are trained to insinuate themselves in a place for a long time, and whose measure of success is based on acquiring intimate knowledge through close observation of the everyday, serendipitous discoveries, and arriving at conditions of "thick description." Who does one choose to observe thickly when the relevant actors include a UN official in an office in Geneva and a pastoralist eking an existence in the margins of the Namibian desert? If one chooses one or the other, the picture becomes incomplete; if one chooses both, the description becomes diluted.

Faced with the methodological challenge of describing a quasi-global phenomenon, where does one begin? What are the options when it comes to arriving at a nuanced picture of something that consists of many "networks" intersecting at various points with the structures of international law?

One set of possibilities can be found in the literature that has exploded onto the scene that explores the new methods of "digital ethnography." As with key concepts in all new literatures, however, there is nothing even approaching a consensus on what exactly digital ethnography is or of what it consists. Discussions of online method have exploded in part because of connections to marketing, making digital ethnography a "human" approach to behavior analogous to the AI systems that produce targeted advertising (and scads of money). One widely shared assumption in this literature is that to understand what is motivating people (from the marketing perspective what makes them click or buy), a researcher has to look more closely at how they share their lives online. On the internet, one can find publicly available information going back several years that serves as an archive of peoples' lives. From this, we can understand what is motivating their speech and behavior "all without ever having to ask them or interrupt them."[30] It is true that the curated lives that people post online reveal a great deal about their tastes and inclinations; and my access to exchanges in Facebook, WhatsApp groups, LinkedIn profiles, YouTube videos, and other social media platforms has taken me a long way toward understanding the ways that abiding senses of grievance motivate what people say and with whom they interact. This is an approach to online research that has also, more accurately in my view, been termed "netography."[31]

This material can be useful, provided one is constantly aware of its uses in studied self-representation, the masks it presents, in much the same way that it is important to get behind the artifices of everyday life. Gabriella Coleman, in her study of the community of hackers known as Anonymous, makes clear in her methodological reflections that online interaction would not by itself have told her all she needed to know about the group, and she took the next step toward knowing her interlocutors by attending their meetings and interviewing them in person, f2f.[32] This confirms the sense held by some researchers working with online communities that the digital-physical binary is more fraught in discourse than it is in human experience.[33] Certainly, there are things people reveal in online space that lend themselves to the classic approach of being open to the concerns of the people studied and giving priority to *not* being directive.[34] But to fully understand what motivates people online, one has to go offline, away from their curated selves, meet them in the real world, and participate, to the extent possible, in the meaning-making, routines, celebrations, and conflicts that give their lives purpose.[35]

Consistent with this observation, what I attempt in this book is to get a sense of what human rights–oriented activists are up to from the ground up, what kinds of thoughts and methods are reflected in their efforts to use new media, and what kind success they achieve (or fail to achieve) through these efforts. In approaching this multifaceted question, I have not only been oriented toward the scholarly and corporate worlds of tech experts—those who are part of tech-savvy NGOs, organized advocacy groups, or the "tech left" or are software engineers with a cause and a conscience—but have been just as interested in those on the margins, the victims of violence with unbroken oral pathways for the transmission of knowledge, grief, and grievance. What brings these various kinds of actor together is their attachment to universal ideals of rights and their public outreach using media technologies in strategic efforts to communicate and leverage those ideals.

Why Despair?

Before moving on to chapter 1, there is a question of terminology that I want to address, which has implications for the approach to intellectual and activist engagement that I take in this book. If, as I argue at the beginning of this introduction, the tensions between hope and despair are manifest in a dialectic, then why have I chosen to give priority to the word *despair*? Why not give *hope* a place of equal standing and possibility? While there is plenty to despair about in the political realm, it might be useful to explore these questions a bit more deeply.

One reason has to do with the fact that imagery is now much more commonly used as a strategic tool of persuasion, and in these terms, despair is far more compelling than hope. A child's lifeless, broken body evokes more intense emotion and attention than one in a high chair beaming over a bowl of cereal, even though both are classic tropes of strategic persuasion. The communications engaged in by human rights activists tend to be constructed around the worst acts and experiences of human life and death. Indignation and compassion accrue especially to those suffering intensely, innocently, and illegitimately, as portrayed through universal symbols that reach deeply into human emotion. Only by cultivating sympathy through portrayal of the lives and deaths of victims can one have any hope of redress. This eventually produces a number of problems, including "compassion fatigue" and skepticism toward the truth-value of victim narratives, which sets in with the constant flow of horrors thrust into our ears and retinas. It can be difficult to escape from a kind of insidious pessimism that follows from the accumulation of acts of persuasion based on what Arthur and Joan Kleinman refer to as the "dismay of images" and the distress of victims.[36]

The other reason for prioritizing the word *despair* follows from a situation in which, rightly or wrongly, the balance of the dialectic has been thrown in the direction of paralytic apathy by prevailing notions of power. In social and political theory since the sixties, there have been two basic approaches to the issue of public participation in political consent and awareness of institutional power, which separate along the lines of their responses to an interrelated set of key questions: To what extent are the forces that shape political consent knowable to the critical analyst? Is it even possible to be critical, and if so, how is it done, what is the method? Are the structures of meaning that contribute to obedience formed at a remove from human knowledge, will, and strategic agency, and if so, how far? Above all, are they subject to construction and manipulation, including in strategies of resistance to "authoritarian liberalism"?

In one kind of response to these questions—prevalent in the United States in particular—scholars have made ample use of Foucauldian notions of power, the idea of power as an impersonal force that can be neither thoroughly investigated nor deflected from its trajectory of domination, taking the form of powerful energies of social control and homogenization. The concept of hegemony that came into wide use in this context is associated with new forms of transnational governmentality, with the spread of NGOs, agencies of global governance, and projects of compassionate corporatism, all of which reach into human souls. In this totalizing way of conceiving power, we are all prisoners of freedom.

And if this is how power is understood, what is the point of doing anything about its abuses? Why waste one's time and energy struggling, when the very act of resistance entails compromise and co-optation? Activism, in these terms, means collaboration with an unseen, unknown enemy. The pendulum of the dialectics of despair becomes unhinged, and a main current of social analysis falls into a condition of apathetic subjection to conditions as they are.

One of the interesting things that has appeared alongside the new strategies and priorities of activism that accompanied the emergence of new ITs is that in the scholarly world there is less acceptance than there was in the 1980s and 1990s of the (largely implicit) conception of power as omnipresent, nebulous, and all but impossible to identify. There is at the same time more effort toward refinement of the methods and concepts for identification of power's forms, strategies, and realms of influence. In particular, less credence is now given to the view that critical scholarship should keep the politics of the present at arm's length. Once one arrives at the idea that there are conditions that repress and others that promote the growth and dynamism of productive critical engagement (however defined), it becomes difficult to stand at a remove as an intellectual spectator. The task becomes to identify those conditions of criticism, cultivate them, and defend them against those who for political ends would undo them.

CHAPTER 1

Street Justice

These suit and ties got the nerve to call it vandalism
They hella mad, say my art is really bad for business
But I'mma paint a better world until the cans are empty

Macklemore & Ryan Lewis, "Buckshot"

Walls That Speak

As I recently walked along a path next to Parc Lafontaine in Montreal, my attention was caught by the words "IS NOT STRAIGT [*sic*]" written on the pavement in red block letters with an arrow indicating the dividing line between pedestrians and cyclists. A quick glance along the path confirmed that even though the yellow line was painted more or less in the middle of the pathway, it wavered to and fro with little precision. Despite this confirmation of the painted words' truth-value, the motivation behind them was puzzling. Were they a lighthearted observation of the urban environment, a reference, perhaps, to the prominence of Montreal's nearby "gay village," using the wavering line as a metaphor for the meanderings of sexual identity? Maybe they were a straight-up sardonic commentary on the condition of the road infrastructure in Montreal, with its potholes and decay, which is a steady source of public scandal, ultimately connected to wider issues of graft and corruption. It "is not straight" also offers a reflection of the nature of injustice, involving something out of order, out of alignment in the configuration of human relationships, something that invokes a sense of discomfort or indignation, a feeling of wanting to "set things to right." The first step toward this correction is often simply drawing attention to the problem, to fact that something in our shared world is not straight, involving others in the sense of injustice, and strategizing toward remedy.

In this chapter I temporarily leave to one side the race for discovery and control of new media technologies and focus my attention in the opposite direction along the spectrum of technological complexity, toward a form of communication that goes back to some of the earliest expressions of life and love from "common" people: graffiti (sing. *graffito*, from the Italian, "a scratch"). At the same time, I am looking for ways that new technologies are intersecting with and reshaping other forms of public expression and outreach. Our attention to intrinsically compelling new inventions could well be distracting us from the myriad ways that human rights claims are communicated and combined with other expressions of grievance, sometimes making use of technologies that go back to ancient wall writing.[1] The key to understanding communication in the current era of human rights is not in an exclusive focus on sophisticated, expert-driven forms of data management but in considering how ITs are interacting with other forms of media and how human rights are combined with other kinds of claim to produce new avenues of expression, public sympathy, redress of grievances, and sources of the self.

As I understand it, graffiti is a form of communication in public space in which the message is amplified by illegality. It can therefore include not only the more familiar spray-painted messages but some forms of street art (including that of its most famous proponent, Banksy) and, more prolifically, sticker and stencil graffiti. As a basic form of communication, illicit messaging in public space goes back to the ancient world, with curses, magical formulae, political slogans, and professions of love among the signs that have reached across the millennia. But there is also something new that is happening in this medium. Street justice claims in the form of repeatable messages in stencil graffiti and sticker graffiti can sometimes be seen as efforts to acquire visibility online. Activists are increasingly making use of links between urban public space and online space as vehicles for affirming collective identity and reaching the public audiences that traverse this space, as a way to "set things straight" or "set things to right." Graffiti artists and messengers have begun to branch out in their use of media. This includes new forms of graffiti used as a way to draw attention to justice causes and act as a hook to bring viewers to overlooked websites.

Although it has its origins in ancient times, graffiti is highly adaptable, with its artists and writers ready to adopt new methods and messages. Sometimes even those taken-for-granted technologies that at first glance might seem "simple" are fairly recent developments. For example, according to Wikipedia, the aerosol spray canister—the foundational technology of contemporary graffiti—was invented during World War II as an efficient way for soldiers to defend against malaria-carrying mosquitoes inside tents and vehicles. It quickly found many other uses, including as a way to apply paint, with its own advantages in terms of technique and portability.

My method here is something like a low-intensity, multiyear series of serendipitous encounters with paint and stickers everywhere that I happened to be. Had I been able to more actively seek out the graffiti writers and artists, I would have encountered a great deal of diversity of origins, claims, and political orientations. This is already evident in the messages, whose authors include activists from the Kahnawake Mohawk territory outside Montreal, immigrants in Paris, asylum seekers in Strasbourg (the site of the European Court of Human Rights), and students in Cairo. The examples of outreach I encountered as an attentive member of the authors' and artists' public give us a glimpse into the rights consciousness (and occasionally its denial) that is so often alluded to as a defining feature of the post–Cold War era.

Conveniently, this seemingly tangible subject matter helps to introduce the somewhat intangible approach to human rights that runs as a thread through this book, one that depends on the imaginations of claimants representing their public selves as they post claims oriented to anonymous publics. This is not a juridical conception of rights, but more an emotional one, based on collective senses of injustice that are actively and strategically promoted. Indignation over the violations, incompleteness, or "crookedness" of law is not only shared within groups on the walls and/or online; it is also communicated to unknown others in vast networks of persuasion and sympathy. Taken together, these acts of persuasion, these public messages of outreach, tell us something about the popular sense of injustice and indignation that is a foundation of the legitimacy and compliance processes of human rights.

My explorations included not only Montreal, Canada, where I live, but also cities where I travelled: in the USA (Boston and Baltimore), Germany (Berlin, Halle), France (Paris, Marseille, Lille, Strasbourg), the Netherlands (The Hague), Switzerland (Geneva), and, once, a bit further afield, in Cairo. In each of these places, I spent time looking for the expression of a violated sense of fairness wherever graffiti could be found: sprayed onto walls in main thoroughfares, alleyways, under bridges, and on sidewalks or stuck to lampposts, railings, balusters, and street signs. My explorations also went online, as graffiti messages—sticker graffiti in particular, with its greater capacity for printed detail—often included URLs to follow for more information on the claim or cause in question. I began actively seeking out expressions of indignation, including what might be called "social justice graffiti," and thinking more deeply about the places they were put and the messages they conveyed.

Of course, not every tag or sticker that might fall under the rubric "graffiti" is about justice claims. I am limiting the phenomena I discuss here to wall art and words, stickers and posters affixed informally and (strictly speaking) illegally with an apparent public message. By this reckoning, flags and campaign posters, displayed with the sanction of the state, would not be included. For the purposes of understanding the communication of injustice as it relates to human rights, I am also not as interested in a broad swath of graffiti that conveys messages unrelated, or only tangentially related, to grievance and injustice, including signatures and gang tags. While still part of the communicative acts and contexts of justice-oriented graffiti, they are quite another phenomenon with different purposes, which I will touch on only now and then. The proportion of all the messages out there connected to ideas of justice is actually very small, depending on the location (as I discuss further below). The cacophony of messages we find on some walls includes advocacy more of anarchy than of anything resembling social justice. Most of the names and symbols painted along railway lines and under bridges are, one way or another, simple expressions of youth identity. Other posts that I came across were commercially oriented, with the product in question revealing quite a bit about the neighborhood in which it was situated: "Sell your house for cash $$$" and "Cash for diabetic test strips" were familiar messages in some of the depressed neighborhoods of Baltimore, whereas on Quincy Street in Cambridge, Massachusetts, a street that divides Harvard Yard from the university's faculty club, a sticker on the reverse side of a street sign reads, "Cash for your Warhol." Besides reflecting the divergent (and widening) socioeconomic status of its target audiences, the mere fact that commercial marketing is now manifested in graffiti is testimony to the medium's potential to reach responsive publics.

Because of their simple, accessible, portable technologies and potential for public response, today's graffiti messages include significant currents of justice claims. Once I became aware of the element of claims making in some of the graffiti where I lived and travelled, I began to see examples of it nearly everywhere I went. The claims I encountered were almost as varied as the possibilities for injustice that provoked them. Just to provide a sense of this variety, here are a few examples.

In the back streets of Marseille, away from the tourist-centric harbor and toward an area with a concentration of seedy cafes and nightclubs, a sticker on a lamppost drew attention to the effects of unprosecuted tax evasion on struggling farmers: "In France, we are 67 million victims of tax evasion. When the government reduces aid to organic farmers for budgetary reasons it is really discouraging!" It includes an effort toward personalization, with an image of a bespectacled young woman holding a hoe standing next to a bale of hay, with the caption: "Alice, 31 years old, organic farmer."

A tag written in differently colored block letters on a black wall in the northern part of rue Saint-Laurent in Montreal reads: "Street harassment is getting old. . . . Get with the times and leave my body alone." In this case, the author has written as though the readers are personally known. The invocation of a failure to "get with the times" implies that the author-artist is aware of a movement of moral progress, a sense that people have rights to bodily integrity that have come into being at least within living memory, rights that are now part of a wider understanding of the meaning and consequences of sexual harassment. At the same time, the author-artist could well be making a wider point, reaching the public that walks along rue Saint-Laurent, which might well be provoked into sympathy by a carefully worded plea.

We should not assume that those communicating justice messages are in step with the main liberal currents of justice advocacy. In Paris, an activist repeatedly sprayed a stenciled message on the sidewalks in some of the main tourist thoroughfares that read, "PARENTS = FATHER MOTHER" (*parents = père mère*), which a web search revealed to be a reference to controversial proposed legislation (since passed) that would replace the words *father* and *mother* with *parent 1* and *parent 2* in the education ministry's official documentation. The intention of the law was to accommodate gay couples with children (*familles homoparentales*), which many on the right perceived as an assault on the traditional family.[2]

Another example proved to be a bit more ambiguous. A sticker pasted to the glass door of the Saint Henri metro station in Montreal announced, "I'm an asshole. I wear fur," which I understood at first glance to be a not-too-subtle reference to a long-active controversy: the decline of Canada's fur industry, particularly the sector based on seal fur, in response to animal rights activism. (It was also, probably not coincidentally, an unapologetic use of English in a province that has legally prioritized the public use of French.) Online, though, things were a bit more complicated. In 2007, the website of the animal rights organization PETA declaimed against the stickers, which were being worn by fur-wearers at after-parties during New York's Fashion Week: "Apparently these 'asshole' stickers have become the must-have accessory of the season for anyone stupid enough to still wear fur."[3] Later, however, users of the discussion website Reddit noted a shift toward the sticker's application to unsuspecting wearers of fur by *opponents* of the industry (categorized under "mildlyinteresting").[4] This shift, in turn, inspired a series of posts in which, for humorous and critical effect, the sticker was applied to cats: "I stick these on my cat all the time. That little bitch wears fur like it's her right."[5] In some cases, then, the meaning of a graffiti message can shift and metamorphose into something quite different from its original meaning—and the only way to see this change is to follow the message online.

I could go on, and in this chapter I will in fact present many other examples of justice-oriented graffiti and its online iterations. Suffice it to say that in my study of the phenomenon, it became clear to me that justice claims in one form or another occupied a significant current of discourse in the anonymous acts of making statements in public space. As shown by the lyrics to Macklemore & Ryan Lewis's song "Buckshot" in the epigraph to this chapter, there is a pervasive will to "paint a better world." My exploration of this phenomenon told me very simply that almost anywhere in the world we might travel, a powerful, wide-reaching sense of justice is there before us, if we care to notice it, quite literally in the writing on the walls.

Illegality

As a medium for justice claims, graffiti has an ambiguous relationship with law. It is often the result of an illegal act. At the same time, it can express a call for justice, a claim for something owed, a reform or fair application of the law. Sometimes it acts as a kind of vigilantism, with the spray can used to apply an immediate sanction relating to a law that is otherwise unlikely to be enforced. In many places its illegal quality is of little consequence, with tags and stickers simply taken for granted in a context of property depression and impunity, but in other circumstances, the illegality of the call for justice is precisely what gives it effect, what brings attention to the message and at the same time marks the illegitimacy of the state, in a way that is impossible in other venues. Members of the public—the graffiti authors' anonymous, unknown passersby—tend to be drawn to illegality, which they use as one point of reference as they perceive and measure the world around them, giving particular attention to acts or artifacts that seem somehow "not right." Hence (though not strictly involving graffiti, but with the same dynamics of persuasion) the technique used by Greenpeace and others of hanging massive banners from buildings, bridges, and monuments, including Mount Rushmore and the Christ the Redeemer statue in Rio de Janeiro. One "stunt" of this kind involved the hotel and golf course in Turnberry, Scotland during an official visit by Donald Trump, in which a banner towed by a powered paraglider read in all caps, "TRUMP WELL BELOW PAR #RESIST." Predictably, the protester was arrested upon landing (probably for trespass), but he had made his point with a global surge of media attention. To make a message stand out by the way it is presented is to bring greater attention and impact to the thoughts it expresses.

My focus on illegality here is not with reference to the graffiti "artists"—those who dedicate their lives to illegal artistic expression in public space, priding themselves on a certain athleticism, assessing risk, running, hiding, and painting in the dark—nor is it on the subcultural connections between graffiti and hip-hop or on the authors of "gang tags," those who mark territory as part of gang activity and sometimes risk their lives trespassing into rival areas with their spray cans.[6] I rely here on what others have said about graffiti artists as members of a subculture.[7] Rather than focus on the members of a single subculture or cohort, I am looking more widely—transnationally—at a phenomenon that does not have one kind of author but has at its base a shared sense that someone's basic rights have been violated and that seeks to convey that sense to others. I pursue my discussion here by giving attention to a marginal subgenre of graffiti, occasional side events to the usual preoccupations of those who put their mark on public space: the calls for justice, expressed in writing (brushed or sprayed), stencil graffiti, and sticker graffiti.

We cannot be certain that these claims are consistently made by a particular segment of society—in fact it could well be entirely misleading to try to identify a "subculture" defined by such things as gender, age, or family income as the source of the writing that makes claims to justice. There is every indication from the messages themselves that the authors vary and include members of gangs, sex workers, and representatives of start-up NGOs—as is clear from the fact that these writings are in a variety of places, using media that vary from the familiar loosely applied spray paint and repeated messages applied with stencils or stickers to the more unusual nail polish or bas-reliefs carved and scratched into walls. My aim here is to focus on the (usually) illegally applied messages themselves: the claims and what they say to me as a reader or consumer and, in turn, what reflections they inspire that might tell us something about the processes of outreach and the hope being found (or sought in vain) in public sympathy.

Graffiti's Publics

Vienna has a reputation as one of the world's historic cities, with its center reflecting the grandeur it once had as former capital of the Hapsburg Empire, but it is also—incongruously—known by a much smaller number of people as one of the world's notable places for graffiti art. I had come here to attend a conference and, during a walk in a free afternoon, found myself following the Danube Canal, with its wide walkways on both sides near the waterline, below the streets and sidewalks that run parallel to it. Here graffiti art and writing flourishes on the walls along some five kilometers through the central part of the city. It was only when a tourist asked me for directions to an official *Wienerwand*, or "Vienna wall" (which I had to admit, I had never heard of and had no idea where to find) that I realized there was an intended design to the apparent chaos of images and writing along the canal.[8] The city, I was to learn, had passed an ordinance designating particular walls along the canal as freely intended for graffiti, indicated by the symbol of a pigeon. Of course, the boundaries between the "legal" walls and the "illegal" spaces next to them are willfully ignored, and graffiti flourishes along the entire length of the canal. All the same, the city's goal has been accomplished, and it rarely intrudes into the historic, conventionally touristic center.

This space gives us an opportunity to reflect further on the idea of the audience, which in this case might include a scattering of graffiti connoisseurs (like the one who asked me for directions). Let us begin with a simple example. The people who wrote "Happy Birthday Anna + Laura ❤" with half-meter-high letters under a bridge along the canal (outside the designated graffiti areas) were in part making a public statement. It is true that Anna and Laura would probably have been overjoyed to come down to the canal and see all the effort their friends put into making such a prominent display of their birthday greetings, and we can easily imagine that they would have felt good to be part of a group of such close friends. This message was clearly meant to go beyond the usual card and cake. But what, more precisely, makes it different? For one thing, the very *publicness* of the display gives it added significance, stemming from the fact that almost everyone walking along the canal will see that Anna and Laura have friends who like them and think of them on their shared birthday, to the point of committing an illegal act to commemorate their friendship. We attach positive emotion to being recognized, not just by friends but—even in the abstract—by strangers.

The same principles of public outreach apply to the justice claims we find in graffiti art and writing, except that the messages and emotions involved are entirely different. The people who wrote NATIVE LAND in one-meter-high red letters, with each letter occupying a window on two of the top floors of a decrepit, abandoned building next to Autoroute 15 in Montreal, were similarly trying to impress their readers by combining a simple message with flouting the city's laws in its execution—in this case, laws against "mischief" and trespass. They were at the same time, with two simple words, drawing attention to indigenous land claims, with the Mohawk communities on the doorsteps of Toronto and Montreal asserting unrecognized claims and with frustration occasionally resulting in clashes (the most prominent being the 1990 siege in Oka, Quebec) between Mohawk claimants and nonindigenous residents.[9] This graffiti in the windows of an abandoned building was a public expression of these unresolved claims. (The fact that subsequently someone broke the windows and eliminated the "native land" message was an expression of resistance to them.) The motorists and their passengers who drove along the autoroute were the subjects of persuasion, of an effort to bring the claim to their attention, in much the same space and with the same basic intention—getting the notice and influencing the ideas of strangers—as the billboards that line the roadway.

This involvement in the opinions of others is key to the claims making that we see painted on the surfaces of urban public space. There is in these words and images a message-in-a-bottle aspect of communication between strangers who are separated by time. Graffiti is often a form of public communication, in which the intended audience is an abstraction. In this sense it is not entirely unlike internet postings, in which the consumers of the communication are often intentionally unknown to the Facebook member, blogger, or web designer, except that writing on a wall limits the audience to those who are in physical proximity to the message or who catch a glimpse of it from a highway or rail line—unless, of course, it is photographed and posted on the web. In some cases, as I will demonstrate in more detail in the next section, there is a blurring of the boundary between wall writings and web postings, with the internet manifesting itself, even recruiting participation, in the mechanisms and messages of physical public space.

Wrongful Conviction

One of the most common themes of justice graffiti centers on cases of wrongful conviction, an especially strong source of indignation. To illustrate this, let us return to the Danube Canal in Vienna, again to a place that is not part of an official *Wienerwand* but some distance away. Here, a few tags next to one another illustrate the occasional legal content of illegal graffiti. One says (in English), "Disrespect my Colors a . . . " The rest is covered by a poster for a concert series, but we can be reasonably sure that it includes the word "and" followed by some sort of threat—a "gang tag," in other words. In fact, this tag is making explicit what many tags do, marking territory, making it known that this part of town is occupied by such and such gang whose colors and claim to territory one had better not violate. This is not so much a public rights claim as a legal marker, a nonstate-sovereignty claim that is the outcome of a collective decision or legislative process, put into writing and reinforced by the threat of force.

Next to it is another kind of legal statement, "Free Fahad!" with Fahad's name underscored. This is a fairly straightforward example of a public justice claim. Fahad, it would seem, is in prison. The author of this tag thinks the imprisonment is wrong, wants Fahad free, and is making this known to a public audience (myself included). The readers are being invited to do anything in their power to bring about Fahad's release, much in the manner of the well-known "Free Leonard Peltier" bumper stickers that display support for a case of wrongful conviction in the USA. That is to say, the phrase "Free Fahad" expresses solidarity with someone in prison, but it is also strategic (however ineffectually, lost in the communicative chaos of a graffiti-filled wall), calling the imprisonment to the attention of a public audience in the hope that something can be done, that the injustice will come to the attention of a public, and in turn (through that public) to those in authority who may have some responsibility for the injustice and/or authority to act to remedy it. "Free Fahad!" is in this sense a public justice claim.

A search using the keywords "free Fahad" uncovers an online movement centered on the case of Syed Fahad Hashmi, who, at the age of twenty-seven was extradited from the United Kingdom to the United States based on charges that he conspired to support a terror group overseas. One article describes Fahad as "a loving, caring, deeply religious man" who grew up and lived in New York and who traveled to London to enter a master's program in political science. Those who support Fahad online argue that his arrest and extradition was unjust, based solely on the testimony of one individual, described as a government-witness "globetrotter," who travelled to England and Canada "to testify against Muslims." An address is provided at the bottom of the web page for sending donations.[10]

The use of English in this justice claim in Vienna provokes another reflection. This was one of many messages along the canal written in English, which seems odd in a German-speaking country. This was probably not so much an indication of the origins of the authors—we can be reasonably sure that few among those out in the streets at night have greater competence in English than they do in German. Rather, it suggests an effort to reach a wider public than the residents of Vienna, to be visible to and legible both online and by the city's many tourists, some who come to see the canal and its *Wand* as a tourist attraction. This is possibly related to the fact that English has become the global language of claims making. Street protesters acting on such varied issues as cartoons depicting the Prophet or the Catalonian claims for independence in the 2017 referendum can be seen online in photographs taken by the global press, acting as intermediaries between those displaying prominent signs in English and their online publics, sometimes in a way that seems at odds with the religious and national sentiments at the origin of their grievances.

Location, Locution

It should come as no surprise that justice-oriented graffiti tends to concentrate in those areas where justice claims are most active and where the aesthetics of public communication are least policed, reflecting populations that tend to be marginalized and underrepresented in formal avenues of redress. Exploring urban public space, it is possible to find locations and landscapes of justice and injustice (and sometimes both at the same time) that draw poignancy and immediacy to the messages of graffiti. A simple example can be found on the street sign for rue Nègre in Marseille, not far from the campus of the Université de Marseille. The first thought that passersby might have (as was the case for me) would very likely be surprise that such an offensive street name should still exist. Then the sticker came to my attention, with a message common to the street campaigns of Europe: "Refugees Welcome." Here, the graffiti was able to draw stronger attention to its message than if it had been randomly placed on a lamppost. It was able to feed off the symbolic violence of the street name and draw simultaneous attention to the claim and the injustice to which it was oriented.

There might be something else going on here. The sticker includes the phrases "Virtus Verona" and "Rude Firm 1921" in the top left and bottom right corners respectively. Virtus Verona is a professional football (or soccer) team, founded in 1921, currently in the third tier (Serie C) of Italian football. Its supporters are widely known in Italy for their pro-immigrant, anti-fascist leanings, as is reflected in the sticker promoting both the club and a "refugees welcome" message. The sticker is therefore at the same time an affirmation of public identity and an example of messaging crossover. The "street marketing" or "guerilla marketing" used by some corporate actors here blends seamlessly with a social justice cause, with the stickers finding their way, through the club's fans, into what would seem to be a Europe-wide campaign.

Perhaps the clearest example of a connection between justice-oriented graffiti and location can be found in the signs posted and stencils sprayed adjacent to the European Court of Human Rights in Strasbourg, France. On the smooth granite paving stones of the tram stop adjacent to the main gate of the court, Reporters sans Frontières (Reporters Without Borders) had spray-stenciled the images of journalists held in Turkish prisons, with the slogan (minus a hashtag, which is difficult to stencil) in block capitals "SAVETURKISHJOURNALISTS." On the railing of a nearby bridge Amnesty International had affixed several posters drawing attention to a case before the court involving one of its own. The poster reads: "Taner Kiliç, honorary president of Amnesty International in Turkey, has been in prison for one year. Because he struggles for human rights in his country, he has been imprisoned under false accusations. One year of injustice. 365 days too many. Demand his freedom at Amnesty.fr #FreeTaner." A poster-sized image of Taner Kiliç wearing a tie and gazing at the viewer with a half smile is attached to the railing next to these words, giving them a visual reference point and more personal meaning. Other information postings were hung with plastic ties on the fence next to the tram station, less professional, perhaps, than Amnesty International's, but following the same script, with information on specific cases being heard or considered for admissibility by the court, often accompanied by photographs of the claimant. One of these included the plea, "Please help United Nation [sic] I am strandled [sic] and I am waiting for the decission [sic] of human rights and Roumanie to finish my case: dossier 1780/10" (see figure 1). Additional information on this case was attached in a clear plastic document folder. The language of this plea tells us of the disadvantage faced by the claimant in his struggle for access to justice. His mistaken understanding of the European Court of Human Rights as an agency of the United Nations reveals a limited knowledge of the law, something that could not be overcome without legal representation, which he clearly did not have.

A lamppost festooned with messages on plastic-protected green paper and more posters spilling out along the fence next to the tram stop outlined the claims of a British citizen, Reginald Sethwick. Reading these posters, we learn that Mr. Sethwick had been subjected to involuntary psychiatric intervention and that his claims had gone nowhere in British courts and were now rejected by

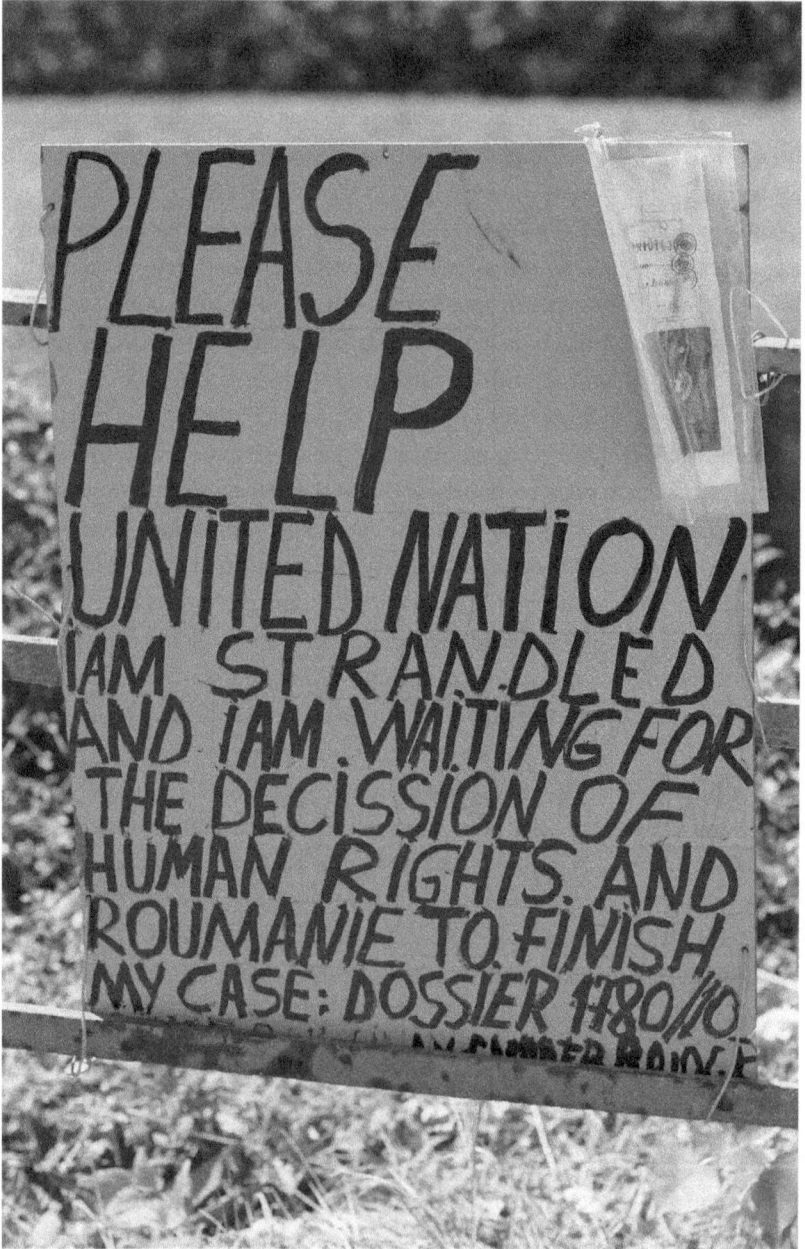

Figure 1 A plea for justice outside the European Court of Human Rights, Strasbourg.

the European Court of Human Rights on the grounds that the application had not been properly submitted. Mr. Sethwick was living in a tent under a bridge next to the court, directly underneath some of his posters. He spent much of his day talking to passersby, explaining to them, to the limit of their patience, the indignities and injustices he had suffered and the unwillingness of the court to hear his case. This was the only instance I encountered in which a graffiti artist–claimant stood by his work ready to supplement it with a personal account of his justice claims, a form of street testimony that seemed to compensate in some way for his lack of access to the court.

If we recognize that justice claims are, in effect, being argued on the streets of Strasbourg outside the court building where cases are being heard and decisions made, then we should be looking more closely at the extent of these informal efforts at persuasion and their connection to place. The only real difference between these public pleas and the social justice graffiti we might find elsewhere in streets and alleyways is the specificity of both the claim and its audience, which no doubt includes court officials, lawyers, and judges on their way to work. The signs and symbols of social justice graffiti in other contexts are usually more general in their messages and are pitched to wider, more random publics.

It might be already clear that a distinct advantage of graffiti is its ability to directly connect a justice claim with a symbolically resonant place, but there is another way that place has an influence on the act of communication: as a way of bringing people with different backgrounds, life stories, ideas, and commitments into contact with one another, with graffiti situated at their points of intersection.

A striking example of graffiti's connection to this kind of intersectional place is a poster-size sticker affixed to a wall near the Quai de Valmy in northern Paris depicting what appear to be illegal immigrants from sub-Saharan Africa crossing the Mediterranean in an inflatable raft (figure 2). Beneath the poster are the words "le droit du sol" (lit. the right of soil), which has to do with the citizenship rights of any child born in France, even to parents without French nationality. Further below are the words "On fait quoi?" (What are we to do?). This graffiti wall is connected to the context of the Quai de Valmy in several ways. First, the area was the site of a temporary reception center for migrants, erected in October 2016 in response to a massive influx of refugees in a collaborative effort between the Office français d'immigration et intégration and the NGO Emmaüs Solidarité, oriented toward "people and families that are the most fragile, the most desocialized, and the most wounded by life: those who live on the street."[11] The center, widely known as the "bubble" (*bulle*) from its inflatable PVC construction, processed more than 15,000 immigrants in 2017 alone. Naturally enough, as a focal point of immigration and homelessness, the site is also marked by a heavy police presence. But the police are not there only for the homeless immigrants. The Quai de Valmy is the site of an active club scene, with the alternative club Point Ephémère hosting both an active nightlife and a thriving graffiti culture, evident on all of its outside walls, with stickers clustered on signposts, gutters, and lampposts—anywhere within reach that the adhesive would work.

The "droit du sol" graffiti wall is thus at a point of intersection between several distinct audiences: the migrants, the club goers, the police, local residents, and a scattering of tourists. What this diversity means is that places of public messaging are not always directed toward those with common backgrounds, interests, and opinions that align with those of the messenger. The symbols and expressions of rights claimants are points of intersection that communicate a common message to different observers at different times. In semiotic terms, they are indexicals, signs that point to an object in the context in which it occurs, indexing, for example, the immediate presence of refugees, police, violence, or race. In their graphical amplification of particular qualities of being, they express social logics and produce identities. This occurs in spaces where the self is rewritten, negotiated, and celebrated in "privileged vectors of the construction of collective memory."[12] Graffiti expresses without inhibition the fears, hopes, preoccupations and indignations of taggers whose collective identities, there for all to see, are experienced as bruised, marginalized, and stigmatized.[13]

Through the representation of collective selves, there is simultaneous outreach to wider publics. Whether or not those who view them are influenced by these public messages, there is an element of serendipity in the encounters that capture their attention. These are small cumulative collisions with difference that weaken or break down insulating barriers to the ideas and experiences of others. The effectiveness of this kind of messaging occurs through the repetition of these encounters, with members of the public coming across a single posting multiple times in course of their daily routine or, as often also happens, the same sticker affixed or stencil sprayed in different parts of the city—one of the methods of low-budget "guerrilla marketing." This makes encounters with the message happen outside of any particular place but with the advantage that the graffiti produces almost unavoidable infiltrations of images and ideas into public minds.

Figure 2 Graffiti in defense of refugee rights, Quai de Valmy, Paris.

Erasure

In our focus on the spectacular images and poignant messages of graffiti, we should not lose sight of the fact that blank space and the residual markings of erasure are also potent symbols. Why is it, for example, that at the International Criminal Court and the International Court of Justice in The Hague nary a sticker on a signpost can be found within a 500-meter radius of the entrance, while other parts of the city abound in messages? A taxi driver to whom I posed this question had a ready answer: the Dutch are an orderly people and the international courts are in upscale neighborhoods where residents are in-tolerant of anyone marking up their public space. Clearly this intolerance of visual "disorder" and the will to back it up with strict policing is something to consider. The International Criminal Court, in particular, is oriented toward war criminals violating international law, and the outcomes of its cases do not affect the lives of European citizens in an immediate way. The people suffering in the aftermath of the conflicts in which the defendants were involved are distant and struggling with other concerns.

By contrast, outside the International Court of Human Rights in Stras-bourg, which I just described, the streets are less residential and more "public"—tolerant of a cacophonous marketplace of pleas and petitions. Yet, all it would take to clear the space of many of its messages would be a state official with orders and a pair of wire cutters. The European Court of Human Rights, with many of its claimants displaced and desperately poor, is tolerant of their acts of outreach and extends that tolerance to the NGOs advocating on behalf of claimants held behind bars in their countries of origin. It actually does the court a service to show it as accessible and impactful on the lives of ordinary people. The contrast between the public messaging (and its absence) adjacent to these various tribunals is informative. Empty walls are not just a product of a neighborhood's social capital and access to policing but also mean institutional distance from the cares and concerns of the people the court is ultimately serving.

Under authoritarian governments, blank and painted-over walls have a very different symbolic resonance. I found an example of this in Cairo where the government of Abdel Fattah al-Sissi has eroded civil liberties and severely punished expressions of dissent. It has quashed the ebullient forms of public expression that took place during the Egyptian revolution of 2011, the period of mass protests against the previous Morsi presidency, with protesters thronging in and around Tahrir Square in Cairo, and the walls of the square and the neighboring streets—including the walls of and around the American embassy—were richly painted with images and slogans.[14] At one time it was possible to find on these walls overtly political messages, mostly denouncing the regime and the violent repression of dissent by the police and the army.

Today none of these striking images and messages remain, other than in the online archives and books that have since been written about the revolution. The central signs on the walls today are those of erasure, with roughly rolled-on paint used to cover expressions of indignation and outrage. A wall near the American embassy shows just such a hastily covered wall (figure 3). A directional sign sprayed on top of the paint-over communicates the additional message that the reason for the erasure was not aesthetic, that *this* piece of writing with its red spray paint and anodyne message can be tolerated, just not anything critical of the state. The covered walls are a clear sign of repression of dissent and, for those familiar with the political context, something more sinister: they point to the lengths the government is willing to go, beyond the application of a layer of paint, to enforce its ideals of patriotism and obedience, not only by erasing acts of communication but the people behind them.

Figure 3 Graffiti erasure in Cairo.

From Graffiti to Social Media

It should be clear from the examples I have discussed thus far that many of the messages conveyed in street art, tags, stickers, and stencils have connections to social media. The most immediate way that this connection takes place occurs through the photography and posting on Instagram and other social media sites, expanding the public viewership from those who encounter graffiti in situ to the global online public who can view it from virtually anywhere, abstracted from context. While graffiti may look much the same today as it did in the 1960s, it has fundamentally altered both its form and audience through its online presence.

Other connections to social media have less to do with the artist or tagger standing back from the work and wanting to share it with an online network and more to do with strategic campaigning, with hashtags and URLs directing viewers to online sources of information about a claim or cause. A challenge of communicating injustice through graffiti is the limited space available to explain the details of an injustice and capture the public's sympathy in a deeper, more informed way. Recall that some of the petitioners to the European Court of Human Rights included details of their cases in ziplock bags next to their shorter written pleas. These were analog efforts of the same kind, in the absence of the money, social capital, or know-how to construct a web page.

Let me give a further example. An advertisement I once came across in a Montreal metro car for a breath-freshening chewing gum ("Arctic bubbles!") included a photo of a young couple embracing and kissing, with the caption "rapprochements sans risques" (encounters without risk). The implicit functional connection between chewing gum and condom use in this poster of was course meant to draw the viewers' attention, perhaps motivate their spending habits, but what drew my attention even more was the fact that it was partly covered (leaving the faces and touching lips still visible, ostensibly so that viewers could see the outrage) by a fifteen-centimeter square sticker with a red X—the universal symbol for prohibition and eradication—and the words "SALE PUB SEXISTE" (dirty sexist ad). I had earlier seen the same sticker graffiti applied to other ads in the city, targeting images that were clearer in their objectification of women. In this case, though, the "sexism" to which the sticker-wielding protester was objecting seemed to be less about gender stereotyping and more about the commercialization of sex itself. The protester was apparently trying to raise awareness of a particular kind of harm, the way that advertising agencies calibrate the images they use to draw upon and promote sexuality, as an illegitimate objectification of a realm of life that should be off limits to the imaginations and manipulations of "admen."

A web search of the key term "sale pub sexiste" brings up a women's student organization, the Association pour une Solidarité Syndicale Étudiante (with the word *étudiante* given the feminine *e* ending). The acronym ASSÉ sounds identical to the French word *assez*—enough. An online brochure (*dépliant*) explains the Sale Pub Sexiste campaign in the following terms: "Advertising is one of the direct and violent vectors of sexism. Sexualized norms are hammered into our spirits every day. Advertising participates in the construction of female and male genders, real social constraints imposed on individuals as a function of their biological sex."[15] The sticker on the poster in the metro is therefore (as I expected) one of many placed strategically on offending advertisements, part of an organized campaign of public outreach, or "continuing the combat" in solidarity with others.

In this example we therefore see something more than the wall writings of a disempowered underclass. There are indications here of a more coordinated and strategic collective purpose, using tools of mass communication—in this case imaging software, a printer, and paper with adhesive backing. The strategy associated with this technology is the simple one of message repetition, creating a meme or a simple message that communicates easily from mind to mind, to make it impossible for anyone who is sentient and literate to avoid at least some encounter in their daily life with the message being communicated. In other words, the protest targets the *ethics* of advertising by using the *methods* of advertising.

When I was recently in Germany (to make use of the state library for a project on German South West Africa), I noticed that this quality of mass communication had been taken a step further by a sticker on a pedestrian walk signal in Potsdamer Platz, Berlin. (This just happened to be not twenty meters from a monument featuring a three-meter-high remnant of the Berlin Wall, the historic reference point for political graffiti). The sticker reads (in English): "FIGHT FOR YOUR DIGITAL RIGHTS!" with a barely visible URL (www.netzpolitik .org) below. If we take the trouble to look this up online (as the distributors of the sticker no doubt hope we do), we find ourselves at a website that describes its purpose as follows:

> netzpolitik.org is a platform for digital civil liberties. . . . Through netzpolitik.org we describe how politics is changing the Internet through regulation and how the Net is changing policy, publics and everything else. We see ourselves as offering journalism, but are not neutral. Our position is: We are committed to rights of digital freedom and their political implementation.[16]

Here, as with the Sale Pub Sexiste campaign, the connection between the words posted in urban public space and those posted online was direct and, more significantly, *strategic*.[17] This strategy takes the form of what we might refer to as "teasers," simple messages or "memes" in public postings intended to attract online follow-up, in much the same way that Hollywood's advertisers use a short montage of a film to attract audiences to the complete product in the cinema.

If there is a demarcation between the solitary wall writing that draws attention to a cause that affects the writer personally and the justice claims oriented toward mass persuasion, it is in the technology used to convey the message. Something as simple as a stencil or sticker allows the message to be precisely repeated, inhibits spontaneity (other than through the location where it is placed), and by virtue of these qualities is oriented toward a basic strategy of persuasion: the infiltration into public consciousness of a simple message multiplied many times over, a meme.

What unites many of the acts of communication I have considered in this chapter is their high degree of commitment to causes, to the point of willingness to commit illegal acts in communicating them to others, and their often strategic outreach to public consumers of justice causes, connected to networks and points of assembly online, based on shared understandings of universal norms and indignation at their violation. Something in the variety of the messages I have described contributes to what I want to say in the next chapter: the current era of human rights cannot be understood in its entirety by focusing on legally knowledgeable, tech-savvy advocates and activists alone. Only when we consider these elite actors in relation to those on the margins, whose messaging might be less effective, and is certainly less technologically powerful, can we really understand the uses and impacts of new ICTs in the era that I refer to as human rights 3.0.

What I have tried to show through the many examples presented in this chapter is that there is clear continuity between the single, unique pleas for justice written by individuals or small groups and the repeated messages connected to organized campaigns. What we call graffiti is sometimes aspiring toward the same thing as the media outreach of NGOs or political parties, though at a more restricted scale. Both are trying to achieve the same kind of exposure, trying to break through the limits of attention and sympathy, to have some mobilization-inspiring influence on the emotions and opinions of unknown others. The main difference between them is that the graffiti artists' means are in most cases more limited. There are usually fewer resources available to those on the margins who are holding a spray can or marker or fixing a sticker to a lamppost, which partly explains their chosen technologies of communication.

Then again, street writing can also be connected to fairly well-organized campaigns of consciousness-raising. If we follow the simple messages that we encounter on the street through internet searches using their keywords or sometimes directly with a URL provided in the messages, we can often find their online assembly points. The most important source of meaning behind message repetition is not just the justice cause but the community that gathers in support of it. Public justice campaigns are at the same time pathways to more restricted forms of collective belonging.

This underscores the fact that we cannot properly understand much of what we see in the world around us without following it into the internet, a bit like Freud's idea of the manifest content of dreams leading toward the repressed desires of the unconscious. Except that once we find ourselves looking for justice causes online, there is little that is repressed or concealed. Here we find the complete representation of opinion, the sense of collective membership in a campaign or cause, and the amplified messages behind the "teasers" put out on the street. Participation by posting a comment or a blog feeds seamlessly into site membership. To varying degrees, opinion in these social justice sites is controlled. The expression of views and values within a certain bandwidth is supported, rewarded with a sense of belonging. The ultimate sources of this kind of low-tech, mass-produced graffiti (if we can still call it that)—the stickers, stencils, banners, and posters we see in the streets—are the online communities that coordinate their members' efforts, sometimes motivating them to break the law in the pursuit of higher ideals of justice.

That is to say, we need to go offline (and then back on) to find the full range of communicative acts that are relevant to current discussions of the new ecosystem of communication and opinion. In this phenomenon there is no stark digital divide between those with the privilege of access and those without, or between spray cans, stickers, and computers. The tools of public outreach are there for all to use. Starting out on this path to consciousness-raising requires only the wherewithal for a can of paint—and the cover of darkness to use it.

CHAPTER 2

Human Rights 3.0

[The historian's] role is to put in order in its historical setting what we experience piecemeal from day to day, so that in place of sporadic experience, the continuity of events becomes visible. An age that has lost its consciousness of the things that shape its life will know neither where it stands nor, even less, at what it aims.

Sigfried Giedion, *Mechanization Takes Command: A Contribution to Anonymous History*

Builds of an Argument

One of the main arguments about the post–World War II history of human rights goes something like this: it is misleading to consider human rights as having a continuous history that extends back to the Enlightenment and the French Revolution; rather, they begin after World War II and have had two distinct phases, with the dividing line corresponding roughly with the fall of the Berlin Wall and the rise of human rights as a global movement. The legal historian Samuel Moyn is the main protagonist of this approach to the history of human rights. According to Moyn, the establishment of the key human rights treaties of the immediate postwar period took place in a very different context for claiming rights from that which developed under the impetus of popular movements starting in the 1970s.[1] Moyn's account of this history places emphasis on the decline of the mostly pragmatic visions of the postwar era and their replacement by human rights as a paradigm of utopian aspiration. "The drama of human rights," Moyn argues in the opening pages of *The Last Utopia*, "is that they emerged in the 1970s seemingly from nowhere."[2] "Within one decade," he continues, "human rights would begin to be invoked across the developed world and by many more ordinary people than ever before."[3] The ultimate source of this agitation and internationalist citizen advocacy was nothing less than a sudden shift in the moral world, a broad acceptance of the ideal of a better world of dignity and respect, even if based on the slow-to-progress, piecemeal reform inherent in the UN system.

Another way of characterizing this shift is by way of a more general transformation from international law to *transnational law*.[4] According to this approach, not just human rights but international law as a whole underwent a shift from simple state-to-state and domestic/international models of international relations to a conception that took into account the many ways that boundary crossing took place between state and nonstate actors.[5] Phillip Jessup's *Transnational Law* anticipated the growing complexity of norm-producing actors and institutions that could be seen only in faint outline in the immediate postwar period, but the real, tangible shift in this direction actually occurred much later. Zumbansen rightly refers to Jessup's insights as "clairvoyant" and situates the emergence of transnational law in a more globalized, fluid world of "denationalized commercial law" and the growing importance of "border-crossing actors" in transnational relationships.[6] This shift in international law is inextricably connected to globalization, a legal response to the rise of a mobile, networked world—not one that reproduced narrow notions of "legality" but that gave more room for the informal work of "legitimacy," the hallmark of human rights.

While I acknowledge that these developments in international law were important, perhaps even revolutionary in their way, in this book I place the emphasis elsewhere: the new era of human rights that Moyn points to as a utopian project and that can be seen as part of the rise of transnational law was also shaped by innovations in the technologies of information and communication that added currency and immediacy to justice claims. The transition toward human rights as a central reference point for the restless agitation of participatory justice movements was given momentum above all by the advent of instant global mass communication, which made it easier for activists to form networks, access information, and appeal directly to public audiences. With this observation as a starting point, the material I present in this chapter encourages a reversal of the usual relationship between law and its artifacts, from an emphasis on values in the history of human rights to a consideration of the social consequences of new information and communication technologies as first-order phenomena in their own right.

When we consider human rights causes from the perspective of their claimants, it becomes clear that technologies of information and communication have historically acted on (or with) them in very different ways—so different in fact that there were deep discontinuities between the forms of activism in what can be seen as one human rights and media era and the next. Every stage in the history of computing has its own aesthetic, its own vision of the future, and, we might add, its own set of organizational capabilities and ways of acting on social worlds that do not eliminate but redefine preexisting cultural forms.[7] These stages act something like Marshall McLuhan's concept of the *break boundary*—with technological breakthroughs suddenly creating new conditions of human control within an expanded environment of social life and possibility.[8] And these break boundaries of technology offer a historical backdrop from which to consider the methods and consequences of justice-oriented activism based on universal conceptions of the human good.

A technologically facilitated transition from a first to a second phase of human rights, for example, corresponded fairly closely with the geopolitical transformations that were central to the emergence of the contemporary human rights era: the period that corresponds with the fall of the Berlin Wall, the breakup of the Soviet empire, and the shift to human rights of the utopian aspirations that were once directed to the possibilities of socialism—but also and at the same time the rise of the internet as a foundation for justice-oriented activism. This historical transformation can be accounted for not so much by a shift in values among intellectuals or through a vague sense of a global transformation in geopolitical ideology as by connecting human rights more closely to the media technologies and communities through which these rights were advocated. This change of perspective involves paying closer attention to activists' use of new technologies, to what they produced and what happened to them as they navigated new global architectures of rights and media.

Seen from this perspective, there is more than one break boundary in the history of human rights that needs explanation. Many legal historians (Moyn included) assume that the contemporary age of human rights began at some point in the latter decades of the twentieth century (precisely when is open to discussion), but in my view another, possibly more fundamental, transformation in human rights, media, and forms of advocacy has occurred more recently—and in some ways is still unfolding. This transformation was ushered in with the rise of social media, big data, cloud computing, and artificial intelligence (among other things) as legal technologies that change the way that justice and state security are implemented and that serve as tools of justice advocacy. The challenging features of this new era include the weaponization of online information and "fake news" and the exponentially growing power of data mining and processing for both the administration of justice and defense against its abuses. The advantages to human rights activists of new technologies include new forms of global instant mass communication for online networking and information sharing (in essence amplifying the advantages of the Web 1.0) and new possibilities for using online data to produce evidence of human rights abuses.

If pressed to assign a precise time frame for this second transformation, I would say it began its gestation with the security concerns following the 9/11 attacks and the rise of the global "war on terror," a term and an accompanying use of technologies that facilitated the repression of many forms of dissent and, perhaps more significantly, the public acquiescence toward surveillance and the loss of privacy as part of daily life. The real breakthrough, however, begins around 2006 and 2007 with the appearance of the major social media platforms, the mass availability of smartphones, and, at the same time, the ability of tech companies to cross a threshold in the processing power needed to make use of massive and variegated data sets (in large part a secondary outcome of social media and smartphones) for processing through AI.

The best way for me to clarify my argument in this chapter is to systematically go through the three periods in the historical sequence of human rights in the seventy years of the post–World War II era to illustrate the points of radical innovation and interruption and the connections between the strategies of human rights lobbying and the technologies that facilitated them. When we perform this whirlwind tour of the various stages of human rights, it becomes possible to see that the distinct historical builds in relation to communication technologies have fairly clear contours. In ways that Gabriel Tarde would have clearly understood from his perspective as a media analyst in the late nineteenth century,[9] the diverse strategies of justice campaigning have closely followed the technical possibilities of public outreach and persuasion. They have all had the common goal of reaching public audiences in campaigns of "consciousness-raising" to leverage the "disinfectant of sunlight" and the "naming and shaming" power of human rights. The ways that they set about doing this have in very different ways followed the possibilities and limits of the technological means of communication with public audiences potentially moved by compassion and outrage in response to state abuses of power.

Analog Activism

In simple terms, the immediate postwar period through the 1980s is the "analog" era of human rights, in which digital technologies were not yet available as tools for organization and public outreach. Activists inspired by the Universal

Declaration of Human Rights and the Convention on the Prevention and Punishment of the Crime of Genocide in the immediate post-World War II era were limited largely to those on the front lines of the hunt for war criminals and pursuit of their prosecution. Initially at least, these efforts were led not by nongovernmental organizations as much as by loose associations of Holocaust survivors, student groups, and their supporters. There were at this time only a few long-established NGOs with an interest in human rights such as the Red Cross and the Women's International League for Peace and Freedom (created in 1915 in response to the carnage of World War I). In the 1940s and 1950s activist networks were not significantly supported by NGOs and had to be formed though other pathways.

This left the mainstream, journalist-filtered media as the central mouthpiece of human rights claims and dissent. Although left-leaning journals readily carried stories and opinion pieces that addressed activist concerns, this kind of exposure only reinforced opinions that were already shared, preaching to the choir. Their real source of leverage, activists soon realized, was the small handful of widely read (and later watched, with the popularization of television) media outlets; their main challenge was to persuade journalists working for major news outlets to cover their stories.

Serge and Beate Klarsfeld's *Mémoires* provides a remarkably detailed account of this era of justice lobbying, as seen through the lens of their efforts to bring Nazi war criminals to justice.[10] One of the qualities that made the Klarsfelds such an effective justice campaigning team was their division of labor into complementary areas of expertise. Serge Klarsfeld cultivated his legal knowledge and research skills, unearthing the documentary record of war crimes and assembling dossiers on each of the Nazi war criminals on whom they decided to concentrate their efforts. Without the meticulous grounding in the documented facts of Nazi war crimes (facilitated by the Germans' penchant for meticulous record keeping), the activism oriented toward bringing perpetrators to justice would have been subjected to doubt and ultimately amounted to nothing.

Of the two, Beate was the more effective in activist methods, and it is to her recollections of the struggle for attention to the impunity of war criminals during the 1950s and 1960s that I turn to illustrate the analog stage of human rights activism. The central vehicle of Beate Klarsfeld's activist efforts in the immediate postwar period was media attention, getting journalists to write stories in major news outlets, based on her conscious realization that the files implicating former Nazis "would only have impact if I accompanied them with spectacular gestures that the press, eager for sensationalism, would hasten to narrate."[11] This was not a passive exercise but involved *producing* public attention to the issue and *generating* stories that would reach a wide enough audience so that government officials would take notice and the state's judicial and political machinery would be compelled into action. It was not enough for her to publish her own opinion pieces in *Combat* or other left-leaning journals or to distribute handbills (which she also did with some regularity). Judging by her account of her activities, she seems to have clearly recognized that to produce enough public notice to dislodge the dominant, politically conservative media influence of a party in power, she would have to reach the entire spectrum of mainstream media outlets.

But how does one get the attention of journalists in major media outlets? Censorship was clearly not the only barrier to media coverage of an important story; so was journalistic apathy and torpor, the unwillingness to take the professional risks that followed from exposing politically connected criminals. What can one do when silence is a collective response to information likely to disrupt a comfortable status quo and offend the powerful?

Bringing notice to the continued freedom and impunity of Kurt Lischka was just such a challenge, involving a war criminal who had sent people to their deaths from behind a desk, one of the "office murderers" (*Schreibtischmörder*) in postwar Germany who remained in comfortable anonymity. Lischka had first been a head of the Gestapo in Cologne, where he led an operation that resulted in the incarceration of more than 30,000 German Jews following the destruction of their property during the infamous Kristallnacht pogrom of 9–10 November 1938. He was then transferred and became commander of the security service (Sicherheitsdienst, or SD) in Paris during the German Occupation, where he was responsible for the single largest deportation of Jews, the infamous Vel d'Hiv roundup of 16–17 July 1942, which resulted in the arrest and eventual deportation to death camps of some 13,000 Jews. Lischka had been deported to Czechoslovakia in 1947 to answer for war crimes but had been released in 1950 and settled in West Germany, where he lived and pursued his legal career—including as a judge—under his own name for more than twenty-five years. Meanwhile a Paris court had sentenced him in absentia to life in prison. And there things remained, with Germany unwilling to pursue his extradition to France while Lischka lived quietly and undisturbed in Cologne.

The activists pursuing the criminal liability of prominent Nazis had to overcome a common rebuttal to the accusations against them: "I was unimportant, a cog in the machine." Or "I was not personally responsible for the crimes of the Nazi regime." Besides, it was readily pointed out that there were so many Germans caught up in the war that if they were all excluded from serving their country in the postwar period, it would soon become ungovernable. Everything would collapse. And if there's one thing the German public hated, it was the sense of disorder in government and its effects on daily life. The press had to somehow be brought out of its complacency and motivated to cover a story that was uncomfortable, that disrupted the status quo. The particular challenge for the activists led by Serge and Beate Klarsfeld under these circumstances was to make the documentary record come alive, to draw attention to Lischka's war record and comfortable life in Germany, garner wide media attention, and break down the barrier of impunity that was preventing his extradition to France.

One strategy that the activist group executed was to stage a kidnapping, which took place on 22 March 1971. A small group meticulously planned Lischka's abduction. They overpowered him at a tram station and forced him into a car, but the plan failed. Undeterred, the activists used the event of the attempted kidnapping to follow through with the press, calling reporters, first under assumed names as "witnesses," then, as the story built, sending Lischka's dossier to the newspapers likely to follow up, as *Der Spiegel* eventually did with an in-depth story. It was, however, Beate, arrested on charges of attempted kidnapping, assault, and membership in a criminal organization, who fully broke open the story. She used the criminal procedure against her to draw attention to Lischka's continued impunity. As she recalled, this effort began with her interrogation: "Whenever they said 'Lischka', I added 'the commander of the Jewish service of the Reich's Gestapo' and if they didn't enter it into the record I would stop talking."[12] Her arrest drew protests by Holocaust survivors, who gathered at the German embassy in Paris, chained themselves to the gate, and posted stickers (*papillons autocollantes*) that read: "Free B.K., imprison the criminal Nazis." Serge meanwhile expanded the reach of press coverage by sending an open letter via telex to the major newspapers in Israel. Eventually these efforts had the desired result: widespread press attention, first to the illegal act of the attempted kidnapping and then to the circumstances—Lischka's war record and conditions of impunity—that motivated such a desperate act. Although Beate Klarsfeld's arrest and the possibility of a twenty-three-year prison sentence were unintended (she was ultimately given two months, suspended), they were among the risks that the activists accepted from the beginning, the means toward *creating* a story that would be widely reported and result in Lischka's extradition to France or arrest in Germany. Aside from possible prison terms, it was win-win.

The desperation of this strategy highlights the activists' reliance on mainstream media during this period. Without the attention of journalists from the main newspaper and television outlets, their campaign was dead in the water.

There were other ways for the war crimes activists to achieve their goal of journalistic exposure. Serge Klarsfeld's meticulous research and efforts in assembling unimpeachable dossiers provided the foundation for activist efforts. Then Beate Klarsfeld spent hours on the phone, calling people one by one, organizing a group of Holocaust survivors, dubbed "the Klarsfeld group," to march on Lischka's home, some dressed in their death camp uniforms (figure 4), first drawing attention to the unusual sight the Holocaust survivors and from there to the war criminal who prompted the protest, his personal history and the circumstances in which he had lived unrecognized and unpunished for decades.

It took nine years for the cumulative effect of this activism and press coverage to have its desired effect. Lischka was eventually arrested in Cologne and, in February 1980, sentenced to a ten-year prison term.

Figure 4 Protesting Kurt Lischka's impunity: Beate Klarsfeld and survivors of Auschwitz in the streets of Cologne, Germany, 1980. Source: Getty Images. Reprinted with permission.

The conditions I have described for the Klarsfelds' campaigns for the prosecution of war criminals were widely prevalent in the first decades of the post–World War II era. There were of course far-reaching innovations in media during this period. The rise of television as a source of household entertainment added a deeply significant element of common emotion and values to the public consumption of images and information. It also led to an increased power of mainstream media, leading to its designation as the "fourth estate," possessing and actively wielding the capacity for advocacy and framing political issues.

The Eichmann trial of 1961 was a watershed moment in which the power of televised media was applied to the pursuit of justice and public awareness of human rights in the aftermath of mass crime. The chief prosecutor, Gideon Hausner, made the oral testimony of victims central to the case against Adolf Eichmann. Amplifying the impact of this testimony, the American firm Capital Cities Broadcasting Corporation (better known as ABC) obtained the exclusive rights to broadcast the proceedings, giving greater immediacy to the event than the global newspaper coverage.[13] In a marked departure from the emphasis on documentary evidence in the Nuremberg trials, this was "the first case of a trial in the aftermath of war to consider the emotions of victims as central to the practice of justice," which "laid the foundation for a new, still unrealized type of transitional justice that has the potential to address the psychological needs of a society after war."[14] Similar arguments about the effects of televised news could be made for coverage of the events of the civil rights movement and the Vietnam War of the 1960s and 1970s.

Even the activism of the AIDS Coalition to Unleash Power (ACT UP) of the late 1980s relied on a strategy reminiscent of that followed by the Klarsfelds in the immediate postwar period. ACT UP's trademark Silence=Death poster, with a pink triangle set against a black background to evoke the symbol used to identify the *"Homosexuellen"* in Nazi death camps, was inaugurated in a demonstration at New York City's General Post Office on 15 April 1987, the day that tax returns were due. The protesters selected this date and location because, as Douglas Crimp observed, the media "routinely do stories about down-to-the-wire tax return filers" and were sure to be at the post office in numbers.[15]

In the early 1990s ACT UP branched out with the development of DIVA-TV, an acronym that stood for Damned Interfering Video Activist Television, an initiative involving video production for local-access television stations, including footage intended to highlight media bias by showing real-time comparisons between original protest footage and the images broadcast by mainstream TV news channels. This initiative was more conducive to adding humor to the emotions conveyed by protest, and ACT UP can be credited with "lightening up" the range of emotions incited by activists in their appeals to publics. The migration to video also involved the tried and true strategy of disruptive protest combined with media baiting. On 22 January 1991, ACT UP took up the very direct strategy of "evening news zaps," which involved disrupting live broadcasts of the *CBS Evening News with Dan Rather* and the *McNeil/Lehrer News Hour* in protest against the Gulf War and the "lack of main-stream media news coverage on AIDS."[16]

While television added immediacy and emotional impact to events that attracted news coverage, it did not entirely eliminate the journalistic filter behind the choice of subject matter reaching the mass public. The efforts of DIVA-TV in its "evening news zaps" clearly illustrates the need for activists to somehow influence or infiltrate mainstream news coverage, even if it meant hijacking broadcasts through direct action. If anything, the limited number of networks that held broadcast licenses increased activists' reliance on mainstream media for coverage of their causes as a source of political leverage. It was either that or consignment to the limited viewership of local access TV. Television may have increased the reach and emotional impact of the images consumed by public audiences, but it did not effectively change the dynamics of justice lobbying. There was still only limited possibility for activists to release the story on their own. They were strictly limited in their ability to reach the broad-spectrum public necessary to change the political and legal conditions behind the injustice and impunity they were trying to expose and undo.

"Netizen" Activism

It is difficult to say with any precision exactly when a transition took place from one kind of human rights activism to another and exactly what technologies might have facilitated the rise in influence of "netizens"—the tech-savvy activists who use their skills for social good. A few notable events in the history of the internet, however, might offer guidance.

- On 26 March 1976, Queen Elizabeth II hit the Send button on her first email message, which read: "This message to all ARPANET users announces the availability on ARPANET of the Coral 66 compiler provided by the GEC 4080 computer at the Royal Signals and Radar Establishment, Malvern, England." In recognition for his work in making the Queen of

England one of the first heads of state to use email, Peter Kirstein was inducted into the Internet Hall of Fame.[17]

- The World, a company based in Brookline, Massachusetts, became the first commercial provider of dial-up internet access, with the first customer logging on in 1989. In achieving this distinction, The World's president, Barry Shein, had to overcome resistance from government and university installations, which threatened to disrupt the project's operating system, Software, Tool & Die. It was not until Shein received an email from the National Science Foundation granting him "permission" to sell internet access to customers other than "bona fide researchers" that the commercial connectivity initiative was able to launch.[18]

- In June 1992, the MIME standard, the internet protocol for sending multimedia messages, was formally proposed and later approved as a draft standard in 1993 by the Internet Engineering Task Force (IETF). The improved email gateways enabled large media files such as photographs and videos to be sent via email. Nathaniel Bornstein, a researcher for Bell Communications Research (Belcore) and Ned Freed, founder of the internet messaging company Innosoft, had collaborated to create the new Multipurpose Internet Mail Extensions (MIME). This one innovation fundamentally changed the way that billions of people interact with technology and communicate with one another.[19]

- In 1993, the White House and the United Nations went online. This was the same year that the number of users on NSFNET (the backbone computer network of the National Science Foundation) reached more than 2 million users, up from 2,000 at its inauguration in 1985. The burgeoning commercial use of the network prompted the NSF to lead an effort to support a new internet architecture.[20] This was the key development that marks the beginning of the internet era.

The innovations that I'm highlighting here (which constitute what is commonly known as Web 1.0) came together in a relatively short space of time to influence the basic contours of a new era of human rights activism based in new possibilities of instant written communication and digital representation. Let us not be shy in acknowledging the significance of this development: it involved the translation of all existing media into digital data accessible to and manipulatable by computer users—and by the early 1990s computers had already become widely available. Within the limits of data storage (which was already growing exponentially), everything in the media world became computable.[21] From the perspective of activism, the rise of *transnational networking* resulted in an un-precedented reach and influence of connection and collaboration among human rights–oriented advocacy groups in different parts of the world. The new global infrastructure of communication made it possible for dissidents to organize and share information much more efficiently than the student organizations and telephone networks that formed the basis of the Klarsfelds' activism.

There was one particular group, the Zapatistas of Chiapas (a province in southern Mexico), that navigated this new media/justice architecture particularly effectively and influentially, tapping into a remarkable level of global support that acted to prevent violent state suppression of their movement. Writing just a few years after the 1994 outbreak of the rebellion, Harry Cleaver boldly stated that "no catalyst for growth in electronic NGO networks has been more important than the . . . Zapatista rebellion."[22] This movement, he argued, has created a new form of social movement and activism, which gives impetus to previously disparate groups "to mobilize around the rejection of current policies, to rethink institutions and governance, and to develop alternatives to the status quo."[23] This analysis has subsequently been amply borne out by the lasting influence of the movement and its scholarly commentary.[24]

At the time of their rebellion, the peasants of Chiapas were responding to a long history of their lands and resources being stolen, with people displaced and subjected to some of the worst labor conditions in Mexico. As with indigenous peoples in many other parts of the world, state policies were oriented toward assimilation, with their distinct language and traditions denigrated and slated for elimination. The final straw was the moment the North American Free Trade Agreement (NAFTA) was put into effect, which allowed for the privatization and exploitation of communal land. The people of Chiapas prepared for this moment, forming themselves into the Ejército Zapatista de Liberación Nacional (EZLN, the Zapatista Army of National Liberation), arming themselves with weapons ranging from a few Uzi submachine guns to makeshift .22 caliber rifles made out of wood. They chose 1 January 1994, the day that NAFTA went into effect, as the moment to begin the revolt.[25]

As it happened, the Zapatista rebellion came to be less known for its military exploits with limited means and more as a movement that came to define the new possibilities of networked activism, widely emulated by others.[26] Although Subcommandante Marcos, the mysterious balaclava-wearing, pipe-smoking intellectual leader of the Zapatistas, was online within days of the start of the rebellion, his efforts were a largely symbolic focal point of a much wider support network. Almost immediately, what began as a traditional guerrilla insurgency shaped itself into what has variously been called an "information age social netwar"[27] and a digital "war of words, images, imagination, and organization."[28]

One of the effects of the rebellion in Chiapas was that transnational activists, who had been surprised by the sudden success of the uprising, created spaces for communication about the Zapatista rebellion on existing internet forums. The mainstream press was reluctant to publish information directly from the rebels, with the exception of the Mexico City daily *La Jornada*, which maintained coverage that eventually also found its way into online information networks. Most of the information on—and especially *from*—the Zapatistas bypassed the paltry journalistic coverage of the rebellion. Communiqués from Chiapas were hand-carried through military lines, publicly posted in nearby towns, and typed or scanned into digital form and uploaded by onto receptive "networks of solidarity" around the world. Receptive readers would spontaneously forward messages to new sites, sometimes translating Spanish documents to English and other languages along the way. The overall effect was that in a very short space of time, the words of the Zapatistas were diffused through much of the nascent internet.

There were particular nodes or hubs of this information exchange that proved to be particularly effective in raising the profile of the rebellion. NGO representatives with global and regional networks built around human rights, indigenous rights, and prodemocracy issues swarmed into Chiapas, creating a need for new Mexican NGOs to facilitate coordination of their comings and goings, the most important being the Coalition of Non-Governmental Organizations for Peace (CONPAZ) based at the diocese in San Cristóbal. The main distributors of the information that came out of Chiapas were university-based online activists, mainly in the United States and Canada. One of the web pages on Chiapas was developed in 1994 under the title "Ya Basta!" (enough!) by the Swarthmore College Computer Society (SCCS). Justin Paulson, an English major and computer aficionado who established the site, had noted the absence of information on the Zapatistas and asked himself simply, "Why not share with the world everything . . . that I manage to find?"[29] Another significant hub of online Zapatista-related materials was the Advanced Communication Technology Lab based in the University of Texas, Austin, whose affiliates also ran Acción Zapatista and the ZapNet Collective. Other points of information distribution did not have a direct pipeline into Chiapas but forwarded material on the rebellion to diverse campaigns of online advocacy. The Institute for Global Communications, for example, established a number of networks in which the Zapatistas were given a central place, including Peace Net, Eco Net, Women's Net, and Anti-Racism Net. Through these and other internet hubs, information coming directly out of Chiapas from visitors and the mainstream media immediately found a place in transnational activist networks.[30]

As the conflict shifted in the direction of a media war, the Mexican government was caught flat-footed. In the period leading up to the conflict, it had been indifferent about the internet and slow to create an information policy that kept up with new developments. In this new media war, doubtful information was quickly checked by the rebels' supporters, with corrections and counterinformation posted within hours or minutes, an unheard-of speed relative to print, radio, or television. Harry Cleaver, an associate professor of economics at the University of Texas at Austin (now emeritus) and creator of the Chapas95 listserv, noted this debunking effect of online communication in an essay written within the first year of the rebellion: "The exposure of lies within an ongoing 'thread' of discussion in cyberspace emerges right up front where everyone can see it. . . . Wild charges of 'terrorism' (echoes of state propaganda) were dissected and demolished in plain public view."[31]

One significant consequence of the information advantage held by the Zapatistas was protection of the rebel movement from violent suppression. The peasants could have been easily defeated by the Mexican military in an armed struggle, but under circumstances in which the state would normally have moved quickly to quash the rebellion, the movement's internet presence made it impossible to invade the Chiapas region without provoking a global backlash and ultimately risking loss of international investment. By representing their cause online, the peasant rebels and their supporters created a standoff that in effect gave them much of the autonomy they were seeking. As the standoff continued, the structure of global outreach grew, based on email listservs and bulletin boards, accompanied by actions of mass protest such as write-in and fax campaigns to Mexican consulates and the US government urging a nonmilitary response to the conflict.[32]

Soon, the internet provided the Zapatista movement with connections to and relationships with activists in Europe and North America that went beyond indigenous issues, to include such mobilizations as the effort to block NAFTA, protests against the unemployment caused by the Maastricht Treaty (or the Treaty on European Union), Italians gathering in Venice to protest regional separatism, and a conference of media activists in New York.[33] Marcos facilitated much of the EZLN's transnational outreach with messages of support to a variety of anti-neoliberal causes. The widening scale of the movement eventually called for some tangible form of international participation. Meeting this need, a conference was held in Chiapas in 1996 under the somewhat overblown title "Intergalactic Encounter for Humanity and against Neoliberalism." The meeting had an equally ambitious result, with some 3,000 activists from over forty countries attending. A second "intergalactic" meeting was held in Spain the following year, which brought together some 4,000 participants.[34] Such a position of global leadership could not have been predicted when the Zapatistas first took up their armed struggle. As Manuel Castells put it, "The Zapatistas' ability to communicate with the world, and with Mexican society, propelled a local, weak insurgent group to the forefront of world politics."[35]

It is impossible to say with any certainty how and to what extent the networked activism of the Zapatista rebellion influenced others, but it was clearly profound and far-reaching. The international movement of indigenous peoples, for example, departed from the Zapatistas in terms of its acceptance of human rights and the institutional framework of global governance as a source of change, but it certainly took up new media with similar alacrity. Some of the key figures of the global indigenous movement whom I met at meetings of the Working Group on Indigenous Populations in Geneva in the 1990s had been in contact with activists from Chiapas through the internet and some had even travelled to meetings there. The main impressions they came away with from these visits, however, had less to do with media technology and more to do with the kind of observation made by Zeynep Tufekci, an activist in the Arab Spring who had made the trek to Chiapas early in the rebellion: "I found a place without electricity, let alone the internet, ruled by a brutal struggle for survival."[36] The main hubs of activist engagement were elsewhere, in the virtual world of information sharing and networking.

A less apparent effect of internet activism has to do with the connection between rights and identity. The advent of the internet introduced new possibilities for representation and *curation of the self*. It put into the hands of the user the ability to modify the default settings of computers and to creatively manipulate the multiuser space built on the web. Illusions built on photography and cinema could be applied to new models of the world and the answer to the questions "who am I?" and "who are we?" Through multiuser space, the human subject could compensate for the atomization and alienation of social reality through the creation of virtual

communities.[37] As with those in the transnational indigenous peoples' movement who readily took to the internet, it also created space for justice-oriented
representation of threats against communities with secure identities in the real
world.[38] The internet, in other words, mapped a new geography of identity, often
connected to justice causes, which followed from the basic feature of individual
user control of the building blocks of media. Computer literacy seemed to confer
something more than just the advantages of networking through new avenues
of high-speed communication; it at the same time simplified and solidified the
connection between dissent and representation of the self.

One of the important things the Web 1.0 facilitated with its powers of networking and curation of identities was the formation of distinct communities
of claimants that lobbied for the development of new human rights standards,
tailored to their particular circumstances. New human rights standards had their
foundation in what some researchers referred to as "online peer-to-peer community venues" that provided access to networks of like-minded others who shared
similar life stories, experiences, and values.[39] These online communities served
as the foundation for the wider ambitions and activism that energized the human
rights movement in new directions. Specific interest and identity groups of the
new millennium took human rights beyond the by-then established instruments
embodying universal concerns about the human condition (the Universal Declaration, the ICCPR, and the ICESCR) and deeper into a realm in which specific
categories and kinds of humanity declaimed in favor of their own juridical instruments and venues. It is especially telling that some of the most recent products
of human rights standard setting—the Convention on the Rights of Persons with
Disabilities (CRPD, 2006), the Yogyakarta Principles on the Application of Human
Rights Law in Relation to Sexual Orientation and Gender Identity (2006), the Declaration on the Rights of Indigenous Peoples (UNDRIP, 2007), the International
Labour Organization's Convention on Domestic Workers (ILO Convention 189,
2013), and the Declaration on the Rights of Peasants and Other People Working
in Rural Areas (2018)—are each the result of long-term lobbying by people who,
beginning in the 1990s, made effective use of new technologies that helped them
overcome their particular structural conditions of isolation, marginalization, and
powerlessness. Human rights became inextricably connected to newly networked
and empowered forms of identity.

The rise of a transnational community of persons with disabilities illustrates
particularly well the counterisolation effects of online communication. Within the
rubric of "persons with disabilities," of course, there is great variety of experience.

People with mobility impairments, besides the social stigma that attaches to their visible difference, face basic obstacles to the formation of social networks in their physical environments. Those who are blind or visually impaired have their own set of obstacles to social networking, including the technical apparatus and social support needed to navigate online. The common denominators that bring these justice claimants together include not only the broadly shared obstacles to social inclusion but also the stigma directed toward them.

The element of publicly expressed stigma was, of course, not at all absent from the newsgroups, discussion forums, and live chat rooms that they actively formed and in which they participated. Much of the optimism associated with the early years of online networking was deflated by the nefarious and insidious consequences of anonymity in the expression of opinion.[40] Aggressive members and trolls were a particularly troubling side effect of online community formation, which led some analysts to the conclusion that difficulties in establishing trust significantly stood in the way of developing online relationships, including their medical use as a source of "mental health and social support intervention."[41] But this didn't prevent communities oriented toward specific disabilities from springing up in an immense variety of forms and forums, developing the nodal basis of networks that transcended the everyday obstacles of space and derisions of face-to-face interactions.

Without the ability to find like-minded people online, the communities forming around new claims and identities might never have existed. In this second human rights era, we therefore saw groups of rights claimants being formed through the technologies with which they were researching their common conditions of injustice.

Human Rights 3.0

The era, or "build," of justice advocacy and campaigning that I refer to as *human rights 3.0* was facilitated by the rapidly increasing amount of online data ("gushers of data" in the apt words of one deep-learning engineer[42]), the rise of social networks as sites of activist mobilization, the growing sophistication of tools for searching and making use of big data and deep learning, and the application of algorithms and artificial intelligence (AI) to questions of policing, rights compliance, investigation of mass crimes, and (albeit lagging somewhat behind) awareness-raising. Out of this constellation of phenomena, six central features stand out as especially noteworthy: (1) *social media landscapes* with platforms that facilitate lobbying and organization, but also introduce problematic forms of discourse inclined toward oversimplification and incivility; (2) new forms of *technologically facilitated rights violation*, mainly those involving new capacities for censorship, the invasion of privacy, and the misapplication of AI; (3) the trend toward *collapse of the "fourth estate,"* represented mainly by investigative print journalism, as a source of public justice campaigning; (4) emerging forms of *collective resistance* to the misuse of technologies—through the *use* of those same technologies, a significant aspect of which involves experts forming the main bulwark against technologically applied rights violations; (5) one current of this specialized expertise is oriented toward "vernacularizing" technical knowledge, creating space for new grassroots forms of *participatory fact-finding*, which, among other things, are expanding forensic capacities in the investigation of mass crimes; and (6) an enclosed and proprietary current of expertise, involving greater *private sector participation* in human rights initiatives, spearheaded by major technology corporations in philanthropic gestures that draw from, while masking the harms of, surveillance capitalism.[43]

Putting these six features together presents us with the contours of a digital arms race involving a contest between abuses of new technologies of surveillance and invasion of privacy and defenses against these abuses, taking the form of efforts to promote and protect technologically enabled capacities for documenting and publicizing human rights violations. The efforts being taken by some powerful states and their corporate allies to control the structure of internet activism (even while some of these same corporations are involving themselves in human rights initiatives) should temper any exaggerated hope that overcoming digital inequality and accentuating the horizontal, universal nature of the web might lead inevitably to new conditions of human rights compliance and participatory democracy. Let me now outline in a little more detail the main features of this most recent nexus of human rights and technology.

Social media and the "post-truth" crisis. To tech-oriented activists, the early years of social media that followed the simultaneous release of smartphones and major social media platforms like Facebook and Twitter in 2006 and 2007 were full of promise. In some respects (as I discuss more fully in chapters 5 and 6 in my studies of Tuareg and Ovaherero/Nama activist efforts), this early optimism has been given some credence with the ability of activists to create communities and navigate pathways of global outreach around carefully curated symbols of their causes. At the same time, however, a dark side to social media has taken form, thrown itself onto the global stage, and established itself as nothing less than a fundamental challenge to democracy and the rule of law.[44] The alarmist approaches to social media take two basic and interrelated forms: those that emphasize their consequences for mistruth and incivility, focusing in particular on Twitter, and those that emphasize their consequences for enclosure of opinion (the "filter bubble" or "echo chamber" effect), focusing largely on Facebook.

The incivility facilitated by Twitter came out clearly in the case of Tay, an artificial intelligence program engineered by Microsoft that was put online to learn how to tweet by actively interacting with humans. It did not take long for the dark side of Twitter to infiltrate and take control of the program through its interaction with users. After just twenty-four hours, Microsoft ended the experiment because Tay had picked up the profane language of other Twitter users, while expressing a marked affinity for Adolf Hitler and posting Tweets rife with racism and hate at every opportunity.[45] Covering this event for *The Telegraph*, Helena Horton makes the sardonic observation that Microsoft was not entirely to blame for this result. After all, she writes, Tay's "responses are modelled on the ones she gets from humans—but what were they expecting when they introduced an innocent, 'young teen girl' AI to the jokers and weirdos on Twitter?"[46] Sadly, Microsoft's AI experiment seems to have passed the Turing Test, devised in 1950 by Alan Turing, in which the threshold for computer intelligence is the point at which its text-based conversations are indistinguishable from those of humans. During her brief day in the world, Tay was well and truly computer intelligence because she tweeted every bit like the worst of us.

This experiment, combined with the better-known proclivities of Trump toward using Twitter as his platform of choice for messages intended to denigrate and humiliate his adversaries, reveals a serious problem with the platform and a source of disquiet about social media with implications for human rights advocacy: its tendency toward crudity and stereotyping stands in the way of understanding the complex conflicts that are at the root of human rights violations. It becomes difficult, if not impossible, to bring attention to conditions of violence and those responsible for them if public discourse is debased by structures of communication that favor intolerance, vapidity, oversimplification, and misinformation.

Another area of concern about social media focuses on its facilitation of enclosed communities of opinion. It has constructed boundaries around distinct publics and inhibited exchange of information, opinion, and influence between them. Siva Vaidhyanathan, in his book *Antisocial Media*, has established himself as one of the leading voices of alarm concerning the social consequences of Facebook. His main argument is consistent with that of others (Eli Pariser, Cass Sunstein, and Shoshana Zuboff prominent among them) who offer dark assessments of the platform's consequences for democracy. Following from its support of emotional content, Vaidhyanathan argues, Facebook is designed in such a way that it narrows fields of perception and creates "echo chambers" or "filter bubbles" of enclosed belief and conviction. It does this by predictively scoring each item, with each user's preferences emerging over time and reinforced by the platform's selection of content. As he puts it, "Facebook does not want to bother you with much that you have not expressed an interest in. Over time your feed becomes narrower in perspective by virtue of the fact that friends and sites tend to be politically consistent in what they post."[47] The result? "Facebook users are incapable of engaging with each other upon a shared body of accepted truth."[48]

Vaidhyanathan's jeremiad goes further than other expressions of concern in its emphasis on the platform's manipulation of users' emotions, its amplification of content that hits strong emotional registers, whether joy, indignation or, we might add, hate. Facebook, he says, is "explicitly engineered to promote items that generate strong reactions."[49] It measures engagement with the number of clicks, Likes, Shares, and Comments on a given post. It can even measure how fast or slow a viewer scrolls through content, so that pausing on content registers as an indication of interest. This helps advertisers define their "targets" and helps Facebook generate massive amounts of advertising revenue, but at the same time, it reinforces the tendency for the most inflammatory material to travel the farthest and the fastest. This would tend to favor such content as puppies, babies, lifestyle quizzes, and hate speech, but also genocides, mass atrocity, warfare, illegitimate death, and injustice, all of which are also highly emotionally charged.

Following this same logic, we can see social media platforms as having qualities that make them ideal instruments for cultivating the kinds of visceral hatred that are at the root of human rights violations. The major platforms have taken some steps, reluctant and foot-dragging though they are, to delete offensive posts and to ban users who consistently violate the terms of use with expressions of hate and/or harmful falsehoods. But these measures have not been vigorous enough to prevent the use of the social media platform as an ideological weapon in the perpetration of mass atrocity. A damning UN report in 2018 on atrocities committed in Myanmar found that the role of social media in the violence was significant:

> Facebook has been a useful instrument for those seeking to spread hate, in a
> context where, for most users, Facebook is the Internet. Although improved in
> recent months, the response of Facebook has been slow and ineffective. The extent
> to which Facebook posts and messages have led to real-world discrimination and
> violence must be independently and thoroughly examined. The mission regrets
> that Facebook is unable to provide country-specific data about the spread of hate
> speech on its platform, which is imperative to assess the adequacy of its response.[50]

Following publication of this report, the country's top general, Min Aung Hlaing,
was banned from Facebook, along with seventeen other accounts, an Instagram
account, and fifty-two Facebook pages originating from Myanmar's military.
Taken together, these social media accounts had 12 million followers.[51]

The common picture of social media we get from the most alarmist critiques
is that its platforms are the ideal instruments for shaping opinion into the kinds
of intolerance that are at the root of human rights violations. This manipulation
ranges from incitement to genocide to the kind of ordinary incivility that has
become part of the air we breathe, which corrodes the legitimacy of law and
acceptance of the factual material that forms the backbone of legal claims and
democratic process. What is more, the major platforms have been reluctant in
their response to this crisis, defending "free speech" not to serve the interests
of democratic exchange but to protect the integrity of the data streams that are
their sources of profit.

This brings us to another important distinguishing feature of human rights
3.0. One of the most consequential features of the new media environment is
the proliferation of overwhelming masses of information, without regard for
its quality or reliability. The most significant trait of the post-truth era is that
the factual foundations of reality are being manipulated *as a strategy of political
domination*.[52] The cues that usually accompany a news story in a mainstream
outlet that say something about its origin and trustworthiness are largely missing
from social media. This quality of social media favors unreliable news sources,
often disguised as solidly credentialed reporting. This free-for-all ability to produce and promote falsehoods disguised as legitimate news, in turn, leaves the
field open for politically motivated abuse of facts and manipulation of readers'
emotions. The platforms, taken together as a media ecosystem, reinforce the
tendency toward emotion as a substitute for truth.

This has important implications for human rights awareness and compliance.
Even the most egregious episodes of state-sponsored violence can be buried

beneath masses of false information, "fake news," and spin (not to mention AI-produced images and video, or "deepfakes"), much of it put out on social media. A patently false denial by a head of state or spokesperson, often repeated again and again to drive the message home, almost inevitably introduces an element of doubt—and a possibility for naysayers to question the truth of human rights reporting.[53] Beyond the mistruths that are part of any media environment, new capacities for hoaxes and digital manipulation make it almost impossible to tell whether any given report is true and has merit. Under these circumstances, how can anyone make a credible claim of a rights violation? It will be simply denied, the credibility of witnesses subjected to smear campaigns and thrown in doubt, and counterinformation given saturation coverage in news outlets sponsored by the very people or organizations responsible for rights violations.

In response to this problem, human rights advocates have attempted several things. The basic strategy has been to publicly unmask the sources of mistruth, reveal their motives, and throw reliable facts at their falsehoods. This can still produce an information standoff that quickly bores audiences. But, as I discuss further in chapters 3 and 4, sometimes enough beneath-the-surface facts can be amassed to take down a politically motivated misinformation campaign, lending promise to emerging, activist-led standards of truth online.

Technologically advanced capacities for human rights violation. In the new era of human rights, not only has the landscape of information technology changed in fundamental ways, but these changes have been accompanied by new and emerging forms of human rights violation. The "G7 Declaration on Responsible States Behavior in Cyberspace," for example, departed from the usual script of global governance techno-utopianism by explicitly recognizing some of the dangers inherent in digital innovation:

> We are concerned about the risk of escalation and retaliation in cyberspace, including massive denial-of-service attacks, damage to critical infrastructure, or other malicious cyber activity that impairs the use and operation of critical infrastructure that provides services to the public. Such activities could have a destabilizing effect on international peace and security. We stress that the risk of interstate conflict as a result of Information and Communication Technology (ICT) incidents has emerged as a pressing issue for consideration. Furthermore, we are increasingly concerned about cyber-enabled interference in democratic political processes.[54]

The declaration recognizes a new, technologically enhanced reality in the acquisition and exercise of power. While a great deal of attention is paid to violent forms of population control and coercion as the main instruments of human rights abuse—the emotionally wrenching deployment of bullets, bombs, forced hunger, and strategic rape—some states and corporations (sometimes together) are quietly violating rights as they encroach on freedoms of expression through digital technologies. (Much more on this in chapter 3.)

Bruce Schneier, in his book *Click Here to Kill Everybody* (he freely admitted in a talk at the Harvard Law School that the title is clickbait), offers a fascinating overview of the vulnerabilities built into a new technology ecosystem and the ways that criminals, corporations, and states alike are racing to take advantage of them. He refers to this ecosystem as "Internet+" since it includes much more than the networked computers that we readily associate with the internet.[55] Our world, as Schneier describes it, is populated with an increasingly dense mass of "smart" devices, not just things that we recognize as digital like computers, cell phones, and GPS navigation devices, but buildings and infrastructure—things like hospitals, factories, and power plants—are now digitalized and interconnected; and ordinary consumer items like locks, lights, refrigerators, toys, cars, thermometers (to target ads to where people are sick), printers, and cameras are all, or will soon be, digital devices.[56] Everything is becoming a computer in the physical, not just the digital, world. These "ordinary" items, which constitute the Internet of Things (IoT), are not always made with the highest standards of security in mind. They are eminently hackable. Exploited singly or in combination, computers, mobile phones, and devices in the IoT can be deployed as instruments of surveillance, control, sabotage, and in some cases—hypothetically at least—assassination. There is, he points out, ultimately no fail-safe security on any digital device.

Schneier's concerns about security in this new digitally saturated world are mostly limited to the realm of criminality, to what he refers to as the "intelligent malicious adversary."[57] But it doesn't take a great leap of the imagination to see that exploiting the vulnerabilities of digital devices is a form of surveillance violence that can be wielded by states and powerful corporations, perhaps not exactly like firing bullets or dropping bombs, but with consequences that can be catastrophic nevertheless. Ultimately, without noise or show, it can result in dissidents getting the proverbial knock on the door. While states have long engaged in the use of cutting-edge technologies for eavesdropping, censorship, and coercion (the former East Germany is a prime example), the scope and scale of the potential for abuse of these technologies today is unprecedented.

The possibilities for technologically empowered counterintelligence point to a more concrete general quality of human rights 3.0: ambitious states have a vested interest in gaining the upper hand in the race for digital technological supremacy, and in the exercise of that ambition, they have empowered themselves as never before to violate the basic human and civil rights of their citizens.

Arguably the most prominent example of a state's readiness to use new technologies to suppress dissidence comes from China. In February 2014, President Xi Jinping announced a policy of internet control to the first meeting of the Central Leading Group for Cyberspace Affairs: "Without cybersecurity, there is no national security; without informatization, there is no modernization."[58] The key feature of China's internet policy is the "Great Firewall," which blocks the country's approximately 750 million internet users from accessing platforms and internet flow from outside its borders. Since 2015 China has upgraded its firewall and suppressed virtual private networks (VPNs) that allow users to circumvent it.

With the passage of a counterterrorism law and a cybersecurity law (both in 2016), the Chinese government gave itself sweeping powers to engage in surveillance and censorship practices, all under the guise of legality and security interests. In practice, the state's interpretation of "illegal" use of the internet (often resulting in jail sentences) has included citizens' disputes with local governments, and its crackdown on dissidence has gone well beyond security interests to target journalists, NGOs, rights defense lawyers (*weiquan* lawyers) and their legal assistants, public opinion leaders, and even the social networks that connect them.[59] An illustration of the global impact of censorship in China was the clampdown on information expressed by Dr. Li Wenliang, who sent out an early warning about the coronavirus from Wuhan Central Hospital and later died of the disease. Ya-Wen Lei, in a comprehensive and meticulously researched overview of China's cybersecurity measures, shows that the Chinese state is particularly reactive to the capacity of internet-empowered individuals and organizations to turn "sensitive events" into political embarrassments that undermine the government's credibility and public image.[60]

Arrests, imprisonment, and public confessions on state-run television are the most obvious repressive measures that follow from the clampdown on unwanted opinion; these are (as of 2015) accompanied by the construction of opinion through some 280,000 Weibo accounts and 100,000 WeChat accounts operated by the government. There has also been a nationwide enlistment of "cyber civilization communication volunteers," who are given monthly quotas for blogs and articles that express the core values of socialism and contribute to building a "clean space" online. This could well include the practice of *gang stalking*, ramping up the kind of shaming and ostracism already common in social media in order to make a person constantly aware that he or she is being simultaneously socially surveilled, harassed, and rejected.[61]

Even with these multiple strategies of control, the state has not put an end to dissent. China's upgrading of techniques of surveillance and censorship, as Ya-Wen Lei points out, has not eliminated public opinion crises or fully silenced critical voices, including those that draw attention to government repression. But by concentrating its repressive efforts on lawyers and opinion leaders and by asserting the interests of cyber-sovereignty in response to international criticism, the Chinese government has severely undermined the country's organized capacities for the expression of grievance and dissenting opinion.

China's control of online opinion has occasionally involved collaboration from non-Chinese tech firms. Arguably the most significant episode of such collaborative repression was made public in August 2018, when a report from *The Intercept* revealed that Google was developing a search engine codenamed Dragonfly that was designed to meet the approval of authorities in the Chinese Communist Party.[62] Google was setting out to secure this approval by blocking content on an extensive list of topics, including political dissidence, free speech, human rights, air quality, and Xi Jinping. The story emerged only after several engineers left the company, citing its violation of the famous "don't be evil" policy. Their main concerns stemmed not only from the search engine's built-in mechanisms of censorship but also (perhaps above all) from the fact that prototypes of Dragonfly linked searches to apps on the users' mobile phones, making it possible for individuals who had searched banned topics to be easily tracked and potentially interrogated and subject to detention. What is more, Chinese engineers participating in the joint venture with Google would have had the ability to update the search term blacklists, while making it impossible for Google technicians to maintain control over state-sanctioned censorship. A follow-up story in the *New York Times* pointed out that Google engineers involved in reviewing Dragonfly had signed off on sections of code without knowing anything about the context of the project and its privacy implications. Jack Poulson, one of Google's senior research scientists, pressed the US Congress for legislation that would prevent such violations of users' privacy. He eloquently expressed his concerns in his resignation letter: "I view our intent to capitulate to censorship and surveillance demands in exchange for access to the Chinese market as a forfeiture of our values."[63]

The coverage given to this story in August 2018 seemed to take senior Google executives by surprise. In the first six weeks after the story broke, the company refused to engage with human rights groups (Human Rights Watch foremost among them), ignored questions from journalists, and rebuffed inquiries from US senators. It was only when it started losing some of its much-coveted engineers in highly publicized resignations that the company responded to criticism by forming a team dedicated to privacy and data protection, releasing a framework for privacy legislation, and appearing before a congressional committee in Washington to offer reassurances that it was following its own framework and acting to "account for and mitigate potential harms."[64]

It is not clear that we can always count on such assurances. The vastly expanded reach of exploitable devices in our digitally saturated world offers new means for repressive states to control dissidence through censorship and invasion of privacy, quite possibly coupled with the use of AI-targeted policing. This is a doomsday scenario for those who adhere to the values (however defined) of democracy and freedom of expression. As states acquire the capacity to listen in on private conversations and hack into computers to raid private caches of data, how do journalists and human rights activists keep themselves and their information safe?

Solutions to this problem involve a shift in advocacy from bringing attention to rights violations (including arbitrary imprisonment of journalists) to exposure of the falsehoods used to cover up those violations. Reporters Without Borders (Reporters Sans Frontières, or RSF), for example, has started a program intended to bring public attention to journalism that serves political or corporate interests, with the intention of fostering wider discussion about media independence. This effort takes RSF beyond its usual mandate of protecting journalists from censorship, intimidation, and imprisonment, and toward a campaign of bringing attention to the most egregiously biased journalism sponsored by states and industries. In human rights 3.0, there is a wider mandate in rights advocacy, which includes not only crimes against humanity but also strategic violations of truth.

In human rights 3.0, the "gushers" of data and unprecedented computing power for processing it have made it possible for engineers to create artificial intelligence based on "deep learning"—that is, digital neural networks in which computers can learn from data the way that babies learn from the environment around them, starting with little knowledge and then acquiring proficiency and familiarity as they interact with new environments.[65] Deep learning, machine learning, and other disruptive technologies of automated data processing pose different kinds of risks to rights-based societies, often through initiatives intended to make justice more efficient. While these deep-learning technologies are still nascent and have yet to be broadly applied, existing uses of AI in new private-public sector initiatives point to conditions and crises soon to be faced.

There are now many state-sponsored programs governed by predictive algorithms, including in law enforcement and the administration of justice. The vast majority of these applications involve supervised learning, in which AI systems are trained to process massive amounts of labeled data, often to identify fairly straightforward things like "cancer"/"not-cancer" or "criminal"/"not-criminal."[66] These data sets are owned by corporations like Google, Amazon, Facebook, and Apple, and by extension the uses of AI are proprietary, based on collections of code owned by major corporations. For that reason, they cannot be subject to careful scrutiny or interrogation. A recent report on artificial intelligence and human rights by the Berkman Klein Center for Internet & Society at Harvard University explores this new area of human rights responsibility and arrives at a cautionary conclusion: "To the extent that an AI accurately replicates past patterns of human decision-making, it will necessarily perpetuate existing social biases as well."[67] Software engineers, with their inclinations toward apposite vulgarity, commonly refer to this problem as "shit in, shit out." (In AI circles and in the literature, it is also referred to as "garbage in, garbage out.")

In criminal law, algorithms are more and more commonly applied to such things as predictive policing, setting bail, sentencing, and parole. At the same time, surveillance technologies are making room for alternatives to carceral infrastructures, in which those on bail or on probation are tracked and sometimes listened in on, a phenomenon commonly referred to as "digital prison" or "e-carceration."[68] Making use of technologically mediated shortcuts in every stage of criminal justice procedure may serve to reduce the overburdened dockets of courts and take away some of the burden of human decision-making, with all its frailty, but at the same time these shortcuts tend to perpetuate and amplify existing states of knowledge, and hence existing conditions of injustice.

Reporters from ProPublica have found an inherent bias in these systems, with defendants of color receiving higher scores and higher bond amounts than those who are white, even white defendants who have a lengthier and more violent offending history.[69] Prevailing prejudices that are built into the data and possibly shared by those designing the technology are entered into algorithms and reflected in their output, which in turn reinforces policy and practice and forms the building blocks of new algorithms applied to other problems, all in a widening cycle of stereotype and oppression. What is more, the assumptions built into code are not plain to see, and lines of code themselves tend to relocate unseen into other algorithms applied to other problems, with the potential for cut-and-paste replications of discrimination and oppression. We readily associate algorithms and AI with innovation and possibility, but by their nature they actually constitute a deeply conservative technology.

How is it ever going to be possible to prevent lines of code from perpetuating and amplifying conditions of oppression in ways that constitute systemic human rights violations? To answer this question, it is important to recognize that there are various kinds of AI applications, each with their own structures of critical inscrutability.[70] The vast majority take the form of *supervised learning*, in which massive amounts of training data (notably the big data of major tech companies) are categorized and applied to a learning algorithm. Training an algorithm to diagnose cancer through exposure to many thousands of images of both "cancer" and "not-cancer" is a well-known example. In criminal procedure, AI is sometimes applied toward predictive modelling—for example, in determining the likelihood a defendant will skip bail or commit another crime—by making complex correlations between data sets like school records, residence patterns, criminal histories of friends, and so on. With this kind of algorithm, exercising rights of due process by making the code of supervised learning algorithms "explainable"—making explicit the data used and the premises used in coding the algorithm—is an obvious place to start.[71] But automated data processing is complex and difficult for those without specialized training (such as judges) to understand, and the problem of explainability almost inevitably runs up against the limits of judicial knowledge.

Then there is the consideration that it is far easier to explain the sequence of decisions used in a simple list-plus-algorithm than for an application based on deep neural networks. How does explainability work when machines learn directly from information coming from their environments, the same way, for example, that children learn languages? This is the *unsupervised learning* now used in emerging technologies like self-driving cars, but conceivably it will be applicable to criminal procedure in the not-too-distant future. In the case of computers bypassing human programmers and writing algorithms as they learn, it is normally impossible for *anyone* to read the code thus produced. With regard to this kind of meta-AI, Terrence Sejnowski, one of the innovators behind the development of deep learning based on cognitive neuroscience, situates the problem of explainability in the model of the brain: "The objection that a neural network is a black box whose conclusions cannot be understood can also be made of brains, and, indeed, there is great variability in the decisions made by individuals given the same data. We really don't know yet how brains draw their inferences from experience."[72] While some might be reassured by the connections between the human brain and the deep neural networks of emerging technologies, it still leads to the troubling conclusion that if problems of accountability have not been definitively resolved in humans, it will likely be even more difficult in the case highly sophisticated computers.[73]

One of the central impediments to making AI accountable stems from a very human problem: the fact that engineers (unlike those giving close attention to conditions of human accountability) have lagged behind in training networks to give explanations as part of their data sets. Kevin Slavin, a research associate at MIT's Media Lab, has noted that the fact that we are writing algorithms that we cannot read "makes this a unique moment in history, in that we are subject to ideas and actions and efforts by a set of physics that have human origins without human comprehension."[74] And even if specialized analysts could comprehend and explain them, algorithms are proprietary. Because of this, tech corporations are legally resistant to exposing code to close scrutiny and thereby compromising the profit motive behind innovation. Returning to the human/computer analogy, the problem of AI transparency is as though human decision-makers were subject to proprietary protections and therefore free to decide as they would, with entire realms of knowledge and its foundations protected and off-limits from being called to account. In short, proprietary AI in the service of justice has created a basic confrontation between a technology-oriented neoliberal order and a rights-based society.

The collapse of the "fourth estate." Earlier in this chapter, I demonstrated that in the analog era of human rights activism the mainstream media was the central reference point of publicity efforts and the main conduit through which opinion was influenced. In a provocative opinion piece in *Wired*, David Samuels makes the argument that the twentieth century's "fourth estate" of influential investigative journalism no longer serves as meaningful check on the power of government (or corporations, for that matter) because its control over the printed and spoken word has been eroded by new media technologies.

> In an age in which every smartphone user has a printing press in their pocket, there is little premium in owning an actual, physical printing press. As a result, the value of "legacy" print brands has plummeted. Where the printed word was once a rare commodity, relative to the sum total of all the words that were written

in manuscript form by someone, today nearly all the words that are being written anywhere are available somewhere online. What's rare, and therefore worth money, are not printed words but fractions of our attention.[75]

This observation, resonant and alarming as it appears, calls for some elaboration. Some media outlets, in point of fact, have *increased* their circulation as many others have collapsed. At the same time, the metrics by which public attention and its profitability are calculated have shifted. In the analog era, attention was seen in terms of numbers of people, as measured by such things as circulation, subscription numbers, and Nielsen ratings. In the age of big tech, the measures of attention are dehumanized, oriented toward such things as clicks, Likes, and time spent. The measures of actions that demonstrate public interest, and hence the value of a journalistic venue, are then represented as deep-learning futures markets—the finely calibrated predictions of future tastes and inclinations—which have become the actual sources of unprecedented floods of media revenue. Even "legacy" journalism, relying as it now does on big tech platforms, is of necessity complicit in the ecosystem of surveillance and the digital market of targeted advertising.

This does not mean that reliable investigative journalism has disappeared from the media ecosystem, but rather that it has become increasingly rare, or in the pejorative terms of populist backlash, more "elite." Those flagship journals like the *New York Times*, the *Washington Post*, the *Financial Times*, and (among the more specialized and adaptive) *Wired* and the social media newswire Storyful that have made the transition to online subscription and maintained readerships and viewerships oriented toward their journalistic values have managed to survive, even thrive in the context of a wider turn away from broadsheet newspapers and quasi-monopoly television networks. This means that coverage in the institutional press is an even more coveted goal of activist efforts at the same time that the possibility of achieving impactful journalistic recognition is more remote.

As we will see in the case studies that I present in chapters 5 and 6, activists are responding to this new media ecosystem by diversifying their technologies and strategies of outreach. Even with Google and Facebook engaged in an unprecedented private-sector centralization of the platforms used to engage in justice lobbying, activists are still making use of the tools that they provide, sometimes to good effect. Not every item in this new media ecosystem is "fake" and not every person navigating it is naïve. The collapse of mainstream journalism represents both an increased threat of surveillance and media control in a superficially "free" free-for-all of untrustworthy information that appeals to base instincts (the techno-pessimist future) and a wider set of possibilities for self-representation that bypasses journalistic filters and allows stories that would once have been overlooked to capture public attention, multiplying the "fractions of attention" made available by social media into potentially mass accumulations of attention and indignation (the techno-optimist future). The reality now is not a victory of one of these scenarios over another but a contest between them, the outcome of which is as yet undecided—and which, given the seemingly limitless human capacity for dissent and going off any kind of rails, will probably never be fully decided.

Tech experts as human rights advocates. Whether or not the doomsday scenario of unaccountable AI being applied to the troika of war, surveillance, and criminal law has actually been realized, what we *are* already seeing is the participation of tech experts in acts, artifacts, and campaigns oriented toward the protection of basic rights. One strand of this advocacy involves technology-based strategies for counteracting the repression and denial of facts by empowering the witnesses of human rights violations and adding credibility to their information (see chapter 3).

The Tor Project is arguably the most significant example of innovation oriented toward protection of internet users from censorship, surveillance, and tracking. Its name is an abbreviation of the term "the onion routing," the idea that internet users' anonymity and security can be secured by a decentralized network in which traffic is routed through multiple servers, a network of relays run by volunteers, and encrypted each step of the way. Onion routing was conceived in the 1990s, but Tor became a globally influential source of digital privacy and rights with the founding of the Tor Project, Inc., as a nonprofit organization in 2006 and the release of the easy-to-use Tor Browser in 2008. This browser became a vital tool of the Arab Spring movement starting in 2010 and was the vehicle by which Edward Snowden made his whistle-blowing revelations in 2013. Such acts of dissidence are reference points for the project, which its developers and caretakers describe as "a labor of love produced by an international community of people devoted to human rights."[76]

Tech experts have become leading figures in upholding international law in the era of human rights 3.0, not just in participatory projects like Tor, but in the justice-oriented commitments of those working for big tech corporations. The specialized expertise needed for technology giants to implement their visions and profitability are not all waiting for artificial intelligence in the service of despotic states to become a more common reality than it already is. Responding to the realization that Project Maven, one of Google's major AI initiatives, was intended for the US Department of Defense, over four thousand Google employees signed a letter to CEO Sundar Pichai stating, "We believe that Google should not be in the business of war."[77] The tech experts' flight from tools of violence also found its way to students, the future of corporate competitiveness and viability. A MoveOn.org petition sponsored by a student group calling itself the Stanford Solidarity Network took a strong position against Maven and other military uses of their skills and demanded transparency for employees concerning the projects to which they were assigned.

To be delivered to Sundar Pichai, CEO of Google

We, students, pledge that we will:

First, do no harm.

Refuse to participate in developing technologies of war: our labor, our expertise, and our lives will not be in the service of destruction.

Refrain from interviewing with Google until it fully withdraws from its contract with the Department of Defense (Project Maven) and fully commits not to develop military technologies in the future, nor to allow the personal data it has collected to be used for military operations.

Abstain from working for technology companies that fail to reject the weaponizing of their technology for military purposes. Instead, push our companies to pledge to neither participate in nor support the development, manufacture, trade or use of autonomous weapons; and to instead support efforts to ban autonomous weapons globally.

106 signatures. NEW goal - We need 200![78]

A similar flare-up of student activism took place at the Massachusetts Institute of Technology (MIT) when, at the opening ceremonies of MIT's College of Computing on 28 February 2019, Henry Kissinger was invited to speak before a panel on AI and ethics. This ill-advised choice of guest speaker, credibly implicated in the prolongation of the Vietnam War and a variety of war crimes during his tenure as secretary of state, as well as a decision to host Saudi crown prince Mohammed Bin Salman at an earlier event, attracted scrutiny to a wide range of questionable sources of funding for the college. An opinion editorial elaborating on the opening ceremonies in MIT's student newspaper, *The Tech*, drew attention to MIT's "growing quest for private sponsorship, military contracts, and the wrong kind of prestige."[79]

Lest we prematurely celebrate the involvement of tech experts with specialized, proprietary skills in causes that support human rights, we should keep in mind that their knowledge, commitments, and aspirations are not always the same as those who might call themselves human rights activists. For the most part, they arrived at their positions through single-minded dedication to the elegance of numbers and the powers of technology. If they have any deep feelings about the possibilities of the future, they are guided by visions of technological supremacy over human frailties—not quite the stuff of dissident suspicion of power and commitments to accountability. Yuval Noah Harari, a best-selling author and "guru" of tech executives, presents a vision of the future in which robots and computers outperform humans in most tasks, including the all-important areas of warfare and economic management. Under these circumstances, he argues, human rights become superfluous: "As the masses lose their economic importance, will the moral argument alone be enough to protect human rights and liberties? Will elites and governments go on valuing every human being even when it pays no economic dividends?"[80] The likely superfluity of human rights has thus become one of the ideological reference points of those who are bringing a technologically dominant world into being.

We do not have to dwell in high-level speculation about the future to arrive at doubts about the tech sector's qualifications to lead efforts toward human rights compliance. As a sector of society or "corporate culture," Silicon Valley is dominated by young men, privileged by the status that comes with the high value placed on their expertise and inclined toward the exercise, if not celebration, of that status. It is no accident that gender diversity in the tech sector lags far behind other occupations, in self-reinforcing cycles of discrimination, harassment, and androcentric hiring preferences.

Women in the tech sector have not passively accepted conditions of denigration, harassment, and unequal conditions of pay and career advancement. Among other initiatives, former venture capitalist Ellen Pao gathered colleagues together to form the advocacy group Project Include (launched in May 2016), oriented toward the goal of using "data and advocacy to accelerate diversity and inclusion solutions in the tech industry."[81]

Yet the diversity problems that Pao and others seek to dislodge are, so it appears, firmly entrenched. An analysis of 2.87 million papers in the computer science literature published through 2018 predicts that, if current trends continue, gender parity is not likely to be reached before the year 2100.[82] Women's contributions to software development are ultimately shaped by the ideas and values attached to source code and disproportionately take the form of activities that are less perceived as "coding."[83] The culture of coding in the tech sector is androcentric, protected by a persistent and powerful myth of meritocracy, with women advanced through the corporate ladder, but commonly moved into "soft" sectors like marketing when they begin to compete with men in high-level realms of technological innovation. These observations invokes challenges that go far beyond the immediate concerns of gender equality in the tech sector, with implications for the *design* of technologies produced in the context of institutionalized misogyny. The people who create and occupy this realm of corporate culture are now among the presumed new custodians of human rights.

Participatory fact-finding. One of the interesting things about human rights 3.0 takes the form of an inherent paradox. New ITs are simultaneously producing different effects at the opposite ends of the spectrum of technical knowledge. Engineers and hackers with unusual insight into the structures of digital technology are the antidote to the misuses of that technology, moving dissidence into a realm in which most of us can only watch from the sidelines—often without even knowing the rules of the game. Meanwhile, people who are making use of the advanced technology include those with little or no literacy skills. The platforms of smartphones are based on icons and other visual cues that make it possible for anyone at any level of formal education to use them, sometimes in surprisingly sophisticated ways. It is mainly for this reason that the use of mobile phones has expanded to the point of producing a new landscape of human rights campaigning and compliance.

As I discuss further in chapter 3, media-saturated landscapes are the loci of new forms of open-source investigation into myriad forms of illegality, including war crimes and human rights violations. And, as with concerns over the digital divide in the era of netizen activism, the capacity for recording and uploading video and images of human rights violations vary enormously. Cutting power and disabling web access is a common precursor of ethnic cleansing. The pillaging of Rohingya villages by the military in Myanmar is visible not from material uploaded to social media by witnesses and victims but by a more impersonal view, the satellite imagery that shows before-and-after images of burned, razed, and evacuated villages.

Tech sector involvement in human rights projects.[84] Shoshana Zuboff, in *The Age of Surveillance Capitalism*, presents what is to date the most ambitious critique of technology corporations, which in essence marks out the contours of a new form of modernity. Surveillance capitalism as she describes it is a form of profit seeking that involves the rapacious collection and exploitation of the unfathomable amounts of data made available by search engines and social media. The carefully calibrated breakdown of privacy by major tech corporations (Google foremost among them) is the source of a "technologically advanced and increasingly inescapable raw-material-extraction operation," the customers of which are "the enterprises that trade in its markets for future behavior."[85]

This new form of capitalism, led by a handful of technology corporations empowered by technological innovation, is for Zuboff a "rogue force driven by novel economic imperatives that disregard social norms and nullify elemental rights associated with individual autonomy that are essential to the very possibility of a democratic society."[86] She assigns to Google the main responsibility for leading the shift from technological discovery to the exploitation of human sources of data: Google, she declaims, "imposed the logic of conquest, defining human experience as free for the taking, available to be rendered as data and claimed as surveillance assets. The company learned to employ a range of rhetorical, political, and technological strategies to obfuscate these processes and their implications."[87]

There is a paradox inherent in the fact that the same big tech companies that are responsible for the systemic erosion of privacy, basic freedoms, and the foundations of democracy are extending their reach into both corporate responsibility for human rights compliance and technology as a tool for its enforcement. In this new global order, states are not the only actors responsible for human rights standards; corporations, particularly in the technology sector, have taken on an increasingly significant place in meeting public expectations of human rights compliance, even though the formal standards of corporate responsibility for human rights have yet to be developed. One of the more obvious reasons for the quickened pace of private sector responsibility for human rights is simple: corporations are the ones building innovative tools that make new forms of human rights violation possible, and only they can build them in such a way that these rights are not trampled underfoot. A more cynical view, supported by Zuboff's analysis, would point to the public relations advantages of corporate engagement in human rights that serves to obfuscate the situation in which these rights and the basic values of democracy are compromised by the very structures of profit seeking on which these corporations are based.

That technology corporations *will* violate human rights norms, possibly in far-reaching ways, cannot be doubted. At the same time, however, their reliance on a limited pool of specialized and unusually gifted employees, many with a strong social conscience, means that they are walking a fine line between seeking profit in unethical projects and maintaining the legitimacy of the motto "don't be evil."

At the 26–28 November 2018 session of the United Nations Forum on Business and Human Rights in the UN headquarters in Geneva, tech companies were prominent among the explicitly identified "hot topics" on the agenda. Aside from new digital technologies, these topics included climate change, the transition to a green economy, and corporate practice in conflict contexts. Much of the participants' attention, however, went to agenda items that had to do with the tech industry: "Disruptive technology: what does human rights due diligence mean for artificial intelligence" (two sessions); "Forum debate: are tech companies a threat to human rights?"; and "Scaling up human rights due diligence through the use of blockchain."

Every argument that points to the responsibility of tech corporations for human rights violations is at the same time an argument for their responsibility for human rights compliance. The specialized nature of the expertise required to build the software tools used for compliance, as well as the proprietary interests involved in developing those technologies, means that UN initiatives in this domain of necessity involve the private sector. Even though their engagements with human rights initiatives constitute only a miniscule financial commitment out of their vast empires, collaborations with the UN, NGOs, and other private sector initiatives offer significant advantages to corporations buffeted by scandals involving collusion with repressive governments and misuse of users' private data. Corporations are able to burnish their public images by developing tools oriented toward human rights compliance, going beyond not doing evil by making what are widely seen as positive contributions to human welfare.

One of the most fruitful ways that tech companies have found to shape their reputations for applying innovation toward the human good has been through engagement with the UN in applying new technologies toward human rights agendas. The Global Pulse initiative sponsored by the UN in collaboration with tech corporations and NGOs hosts a variety of participating media laboratories to provide information intended to facilitate global governance initiatives, including global analysis of Twitter feeds to measure popular opinion on state governance and the UN's Millennium Development Goals.[88] Microsoft, consistent with this example, has joined with Facebook, Twitter, and YouTube to apply its advanced AI processing to design and advance UN initiatives in security and human rights. It formulated a policy paper, *A Digital Geneva Convention to Protect Cyberspace*, intended to create rules of customary international law "to protect the public from nation-state threats."[89] Responding to the shared priorities of states, it started by creating the joint-venture Global Internet Forum to Counter Terrorism, whose vision statement includes the commitment to tackle "extremist and violent content on our platforms."[90] It also applied its technology to a partnership with the Office of the UN Commission of Human Rights intended to develop and make use of advanced information technology to "predict, analyze and respond to critical human rights situations," which are now proliferating and growing in complexity.[91]

In some ways, the UN–big tech collaboration is a match made in heaven. There is in global governance a marked proclivity to build policy and intervention around indicators, and the application of AI backed by unprecedented computing resources produces powerful, similarly technocratic tools for intervention. The computing power of big tech is able to match global ambitions of UN agendas in a way that no organization in civil society can now hope to achieve.

Not everything the UN does requires the application deep-learning algorithms; as I mentioned above, more often the work of humanitarian intervention makes use of new technologies in the forms of supervised learning through lists and algorithms, a technological extension of the UN's near-obsession with indicators.[92] Fleur Johns points out that the pairing of list-plus-algorithm produces a new kind of juridical structure that is being applied on a global scale across a wide range of sectors, including environmental conservation, refugee policy, nuclear nonproliferation, humanitarian aid, and counterterrorism—all based on preexisting priorities and rationalities.[93] Public-to-private transfers of power produce a "vertigo-inducing unknowability in global policy making" in which, Johns points out, the pairing of lists and algorithms wreaks havoc with the ideals and inducements of transparency. There is an inscrutability to algorithms, which derives not only from the proprietary interests built into the "black box" of code, but also in qualities that technicians sometimes refer to as "agility," their mutability and transferability across different areas of intervention.[94] It may be true that "explainability" is a possible benchmark for regulating the application of AI in ways that serve the interests of justice,[95] but complex processing systems are often moving targets, and finding explanations to hold them accountable can be especially onerous when lines of code are transferred from one application to another. As Johns puts it, new uses of IT bring out "the problem of laboring to render one form of lawful authority transparent to some delimited demos, only to find it inextricably entangled with authority produced elsewhere."[96]

In the process of tech corporations' involvement in global governance, a shift has taken place through an expanding role of data mining in the work of human rights. No longer are data "scraping" and analysis merely the technical preoccupations of highly specialized experts; they have become fixtures in practices of global public concern.[97] At the same time that NGOs are empowered with tools of data analysis, a new style of global governance has emerged under the impetus of new technologies, in which inscrutable digital tools are often "bolted together in a piecemeal fashion, customized, reused, and repurposed away from the settings in which they were originally developed."[98] The inscrutability that this use of AI produces, with code overlaid by outward structures of transparency, is entirely in keeping with the goals of global legal institutions. There is power in secrecy, all the more so when no one actually has access to the knowledge being kept secret. States may be able to control the agendas of many global institutions, but they are unable to demand insight into the logic behind new processes of data extraction when there are as yet no structures of explanation to make that logic transparent.

Human rights have been more visibly supported by collaborations of tech corporations and justice-oriented NGOs in ways that, taken together, complicate the story of corporate rapacity presented by Zuboff. A potpourri of human rights–oriented initiatives (though, unaccountably, without once using the term "human rights"), for example, can be found on the website of Google Jigsaw, the title of which is meant to acknowledge, "that the world is a complex puzzle of challenges." Jigsaw describes itself in its vision page as, "an incubator within Alphabet [the parent company of Google] that builds technology to tackle some of the toughest global security challenges facing the world today—from thwarting online censorship to mitigating the threats from digital attacks to countering violent extremism to protecting people from online harassment."[99] These initiatives have included grants from Google Giving to anti-human-trafficking efforts, in which the company contributed a total of $14.5 million in support of a global data-sharing collaboration between the NGOs Polaris Project, Liberty Asia, and La Strada International. Google did not lend its digital-processing capabilities to this initiative but facilitated the collaboration of the NGOs with the tech corporations Palantir Technologies and Salesforce.com.[100]

Google was more directly involved in building the initiative titled Montage, which it then handed off to be maintained by Storyful (described on Wikipedia as a "social media intelligence agency") while running on Google's servers. The central goal of this initiative was to make use of YouTube's massive collection of user-uploaded data into a forensics resource, involving analysis of everything from war zones to protests (and their repression), enabling the extraction of solid evidence of human rights violations or social injustice.[101] Jigsaw president Jared Cohen explained to a reporter for *Newsweek*, "Montage uses technology to help lift the fog of war, giving journalists and organizations the tools to tell rich and accurate stories from conflict zones around the world."[102] This initiative relates back to my earlier discussion of the development of participatory fact-finding, in which war crimes in media-rich environments produce explosions of photographic and video evidence online. One of Google's more significant contributions to human rights has thus been to provide the technological tools for sorting through this data and making it accessible as evidence.

New technologies are producing powerful tools of forensics, cutting through much of the obfuscation associated with "post-truth" and facilitating programs to assist human rights defenders and journalists who are working to expose human rights violations in conditions of state repression. Tech corporations are leading many of these efforts, even as their technologies break down the norms of privacy and compromise the goals of transparency—creating the foundation for new human rights violations.

Viewed more cynically, big tech corporations are using human rights and "human development" initiatives to outsource their ethics, while their central concerns are with corporate growth and profit. Big tech is privately merging with Big Brother, while publicly contributing tools that facilitate activism. There is a kind of schizophrenia to its participation in global governance. These corporations regularly violate rights while acting to defend them, a bit like simultaneously abusing a dog and donating to an animal shelter.

Taking all of these recent innovations and political developments together, it should now be clear that human rights are not what they used to be. Certainly they are no longer the same as they were in the immediate postwar period when the ink was still fresh on the Universal Declaration and the Genocide Convention and when rights-offending states could bask in the conditions of their impunity. The situation has also changed from the time when the internet first made its appearance and technologically empowered activists formed new communities of claimants and situated themselves in world-shrinking, rhizomic networks of activism and information. The still-emerging era of human rights is very different.

In this chapter, I have tried to sketch out the contours that indicate how and in what ways this difference is manifesting, to "put in order and in its historical setting what we experience piecemeal from day to day," as Sigfried Giedion expresses the historian's role in the epigraph to this chapter. Put simply, this order is marked by two very different kinds of advocates involved in resisting and investigating rights violations: those for whom digital technologies are new and suddenly there in the palm of their hand and those for whom they are a career that involves, in one capacity or another, constructing the new architecture of a digital world—a pairing of two forms of knowledge that make up the subject of the next chapter.

CHAPTER 3

Belling the Cat

> The Mice once called a meeting to decide on a plan to free themselves of
> their enemy, the Cat. . . . Many plans were discussed, but none of them was
> thought good enough. At last a very young Mouse got up and said: "I have a
> plan that seems very simple, but I know it will be successful. All we have to
> do is to hang a bell about the Cat's neck. When we hear the bell ringing we
> will know immediately that our enemy is coming." All the Mice were much
> surprised that they had not thought of such a plan before. But in the midst
> of the rejoicing over their good fortune, an old Mouse arose and said: "I
> will say that the plan of the young Mouse is very good. But let me ask one
> question: Who will bell the Cat?"
>
> "The Cat and the Bell," *Aesop for Children*[1]

The New Cat

A still-emerging criterion of authoritarianism has recently come out of the shadows with a brazenness that lends an element of simplicity to political analysis:
ruling classes in many parts of the world have achieved unchecked authority
in proportion to their ability to take control of the thoughts expressed on the
internet. For dissidents, the act of communicating from mind to minds in email,
blogs, or on social media in ways that are critical of the state thus becomes interrupted by the private, uneasy thought that sometimes, somewhere, somehow an
agent of the state might be following one's keystrokes.

In one approach, the regulation of the internet by secret services is done
subtly. The political ferment that can follow from a prominent injustice or aggression by the state can be prevented with a strategic campaign of doubt, not
only in the form of the usual denials by spokespeople but in efforts to sow
public confusion and maintain the loyalty of core followers by infecting social
media with "alternative facts." "Comedy is now diplomacy by other means,"

writes Jonathan Freedland in an essay about Russian media for *The Guardian*: "Defeated on the first hundred facts, the most ardent defenders of the Kremlin narrative will simply move on to the next hundred: their ingenuity in explaining away hard evidence is a bottomless well."[2] Or, as Zeynep Tufekci, a Turkish activist and analyst of new media, eloquently puts it,

> The most effective forms of censorship today involve meddling with trust and attention, not muzzling speech itself. As a result, they don't look much like the old forms of censorship at all. . . . They look like epidemics of disinformation, meant to undercut the credibility of valid information sources. They look like bot-fueled campaigns of trolling and distraction, or piecemeal leaks of hacked materials, meant to swamp the attention of traditional media.[3]

Invoking doubt about basic facts is just one of the strategies behind state-sanctioned digital intervention in many countries around the world. Another consists of the more familiar blunt instruments of censorship and, if necessary, arrest, imprisonment, and torture. In Turkey, the government of Recep Erdoğan perceived a particular threat in Wikipedia as it set about blocking access to web pages and websites and, as the perceived need arose, restricting access to social media platforms. In Egypt, the government of Abdel Fattah al-Sisi passed the Anti-Cyber and Information Technology Crimes Law on 18 August 2018. The law was ostensibly oriented toward national security and the goals of combatting terrorism, but has been applied in arguably the most sweeping regime of internet control in the Middle East.[4] Corresponding with this turn to control of social media, a Lebanese tourist was arrested at the Cairo airport and given an eight-year prison term for a viral Facebook video post that responded to her experience of sexual harassment and referred to Egypt as a "son of a bitch country." When her arrest, conviction, and sentencing went even more viral than the original video, her sentence was reduced to one year suspended and she was quietly released.[5] Wide reporting on a scandal like this, far from causing embarrassment and political crisis, actually serves the authoritarian interests of the state. It insinuates the thought into the mind of every potential or actual dissident: If we can do this to a tourist for an innocent rant, just think what we can do to you.

Saudi Arabia, in its efforts to suppress activism in the interest of maintaining the absolute monarchy, has not hesitated to imprison human rights activists under the pretext of defense against terrorism, even when the protesters' activities involve nothing more than the dissemination of information and expression of opinion. The fate of thirty-year-old human rights advocate, Israa al-Ghomgham, for example, is unknown as I write these lines. She was arrested and sentenced to death in a secretive trial in August 2018, in part for her documentation and social media posts of unrest in the Shia minority region of Qatif. According to a *Newsweek* article republished by Human Rights Watch, "In al-Ghomgham's trial, the Public Prosecution is seeking the death penalty against five of the six activists based on a host of vague charges that include 'participating in protests,' 'attempting to inflame public opinion,' 'filming protests and publishing on social media,' and 'providing moral support to rioters.'"[6]

Let us not forget China and its so-called Great Firewall, a combination of prohibition of access to the world's internet and the development of state-sanctioned and controlled platforms for the internet and social media. A parallel digital structure of state control is the Social Credit System (SCS). Launched at the national level in 2014, the SCS rates individuals and other entities with a social credit score or rank, ostensibly measuring "the trustworthiness of Chinese citizens in keeping their promises and complying with legal rules, moral norms, and professional and ethical standards,"[7] but with uncertainty remaining about the criteria being applied to produce scores through big data analytics. The mere knowledge that the SCS exists reduces individuals to transparent selves and brings into reality with startling clarity a technologically facilitated Orwellian state.

With its system of social credit in place, situated alongside control of social media—and hence of public opinion—China has arguably gone the farthest in terms of repression of human rights activists. "Rights defenders" (*reiquan renshi*), who had once, during the 2000–2013 period of economic expansion, engaged in a full range of civil rights issues—including access to housing, due process, HIV patient rights, and more—collectively produced what Sarah (Meg) Davis refers to as "the brief golden age of China's almost-open society."[8] But with the rise of Xi Jinping to the presidency and his subsequent purge of potentially rival senior Party officials, the conditions for rights defenders has shifted in the direction of deep authoritarianism, with the government acting to disbar, jail, and torture lawyers, legal assistants, and activists engaged in rights defense. Undisguised tracking and following, abductions, and televised "confessions" have become commonly used techniques of intimidation and public influence.[9]

Such repression of expression goes farthest in regions deemed to be sensitive from a security standpoint. In Xinjiang province where China has engaged in a brutal crackdown on the Uighur Muslim minority, there is—alongside a mass detention program that has put anywhere between 200,000 and 2 million people (estimates vary widely) in "reeducation camps"—a parallel digital infrastructure of surveillance and control. This includes expansion of police stations responsible for surveillance of each corner of the province, frequent checkpoints using biometrics to scan irises and fingerprints and check the contents of smartphones, QR codes (matrix barcodes) to track every knife sold in the province, a network of security cameras, and a survey of residents used to rank "trustworthiness" (frequent prayer is a warning sign). Individual identity cards are linked to "machine-learning systems, information from cameras, smartphones, financial and family planning records and even unusual electricity use to generate lists of suspects for detention."[10] Access to the region by human rights advocates is impossible and information originating from it is spotty, making it difficult to verify claims of one of the most significant ongoing cases of mass-scale human rights violations. Such deep authoritarianism was the background to persistent mass demonstrations in Hong Kong, fueled by concerns about legal encroachments aimed at breaking down the protections of the One Country, Two Systems policy and subjecting the Special Administrative Region of Hong Kong to the full power of China's surveillance state.

The examples of such repressive state behavior that I could invoke are endless. But it gets worse. It is also changing form, becoming more brazen, transnationalizing. This is not an issue that merely concerns the "classic" human rights scenario of governments repressing their own citizens. Using the internet's global infrastructure, authoritarian governments are reaching beyond their borders to engage in espionage and cyberattacks on a global scale. Some states have targeted diasporic communities of dissidents with denial-of-service (DoS) attacks—flooding and overloading websites with superfluous requests—and phishing attacks, setting up false emails and login pages to gain access to private information.

Such was the case, for example, with the Tibetan diaspora in Canada, as illustrated by a report compiled and posted online by the Toronto-based digital advocacy organization The Citizen Lab.[11] In January 2018, Tibetan activists, journalists, members of the Tibetan Parliament in exile, and the Central Tibetan Administration were targeted with a phishing and malware campaign, a reprise of similar events that had taken place in 2016, suggesting this was a concerted effort from the same source. The unpatched (using code with known security weaknesses) computer systems being used by the underfunded Tibetan activists made a relatively easy target for the attackers, who were able to use "known exploits"—vulnerabilities in computer systems that are widely known and referenced in databases—open to attack using basic Remote Access Trojans (RATs).

What impressed investigators was not the technology used in these attacks, which was every bit as simple as the obsolete computers they were targeting, but the sophisticated "social engineering" behind the attack, reflecting careful thought about how to lure unsuspecting targets into entering passwords into false entry boxes or opening attachments containing malware.[12] Citizen Lab's investigation reveals in minute detail everything about the operation except its provenance, which it was unable to determine with certainty, although everything it describes points in the direction of China. The custom-made malware, however, clearly originated from what it calls a *closed espionage ecosystem* that includes "intelligence customers" who initiate the activity, developers who write the malware, and operators who conduct the campaigns.

Taken together, these developments point to an alarming shift in state use of new ITs. To use social media shorthand, the world is in a wtf moment. The ways (and technologies by which) states are suppressing dissidence and resisting being called to account for breaches of international law and overreaches of power are unprecedented. And this, almost needless to say, raises the question of what to do about it.

Defending the Defenders

At a seminar titled Software for the Social Good held in October 2018 at the Harvard Law School, a group of tech-savvy legal scholars presented a what's-happening and how-to on the ways that software design projects are contributing to online safety, privacy, secure communication, and understanding between people. The presenters were infectious in their enthusiasm for knowledge of coding to leverage social good and serve as the foundation for people to "advocate, educate, and communicate for a cause." One contributor said that by learning to code and exploring open source projects and open data, "anyone can be a change agent." The word *anyone* in this context was meant to apply broadly to individuals, NGOs, corporations, and governments; but it isn't entirely clear that in a global context the ability to code extends to all those who are human rights claimants subjected to violence and the unlawful will of the powerful. While coding has opened amazing possibilities for many forms of "social good," it remains grounded in a relatively high level of technical expertise. The instrumental knowledge and creativity called for in developing initiatives of, say, internet monitors that process vast amounts of online activity looking for the use of blocking and controls by government censors is simply not available to most activists, least of all those who are the survivors and subjects of mass atrocity.[13]

This technical specialization in human rights advocacy and "defense of the defenders" has produced a new kind of NGO, the digital advocacy organization, which owes its existence to innovative tech space that draws from both the corporate and academic worlds. As authoritarian governments suppress every public form of dissent, their repression of communication among activists and outreach to public audiences most often takes the form of espionage and aggression through digital technologies.

This situation is in some ways analogous to the place of radio operators in occupied Europe during World War II. Among these, Noor Inayat Khan has belatedly emerged as a leading figure. Code named "Madeleine," she was trained by Britain's Special Operations Executive (SOE) when it was discovered that she had the irresistible combination of fluency in French and training as a radio operator in the Women's Auxiliary Air Force. She was sent into occupied France in 1943, just as the PROSPER network of which she had been a part was decimated by the Gestapo. She evaded capture for nearly six months by changing location on a daily basis. During this time, she served as the sole radio operator for the resistance in Paris before being captured, tortured, and executed in Dachau in October 1943.[14] Khan's story is a remarkable illustration of the place of communication in the resistance to dictatorship.[15]

My parents grew up in The Netherlands under the German occupation and would sometimes tell me about the repression of communication from another angle, from the perspective of ordinary people who were not fighting the war as soldiers or spies but living under conditions of the German occupation as civilians. Those who were desperate for news about the progress of the war faced a difficult choice: ignorance was intolerable, but listening to radio broadcasts from what the German authorities referred to as *Feindsender* (enemy broadcasters), mainly the BBC but also Voice of America and Radio Moscow, was punishable by imprisonment or even death. Many chose to risk their lives and tune in to the illicit signals. Later they would tell trusted family members and neighbors about what they had learned, so that news from Allied sources got out to a population desperate for any information relevant to their fate.

This comparison between censorship in occupied Europe and that of authoritarian governments today invites reflection about what is different in the era of human rights 3.0. What is happening today that dissidents didn't experience in earlier times and troubles? It is tempting to say that the repressive apparatus of the state today is not being fully brought to bear on acts of communication the way it was under the Nazis, but that wouldn't be entirely true, not everywhere at least. What is different today, it seems to me, is the extension of struggle between dissidents and authoritarian governments from illicit acts of communication to the *tools* of communication. Germany in World War II could not fully block the radio signals coming from Allied sources because to do so would interfere with its own ability to broadcast and undermine a major strategy of its all-important propaganda ministry. Today, internet censorship can target the specific expressions of unwanted opinion and the unwanted networks of activists. In more ambitious approaches to censorship, it is oriented to the elimination of entire internet platforms, all the while keeping state control of media intact.

The contest between state-sanctioned campaigns of online espionage and sabotage and the work of digital advocacy organizations is an emerging scenario that is shaping what it means to engage in human rights activism. The tip of the wedge of these advocacy efforts is situated in higher education platforms that combine the technical expertise of scholars and students with tools from the private sector and donations from civil society organizations and the public. The Whistle project based at the University of Cambridge (currently in the alpha stage of development), for example, is being designed to cut through the "verification bottleneck" and lend trustworthiness to "civilian witnesses" who capture events and then publish footage and photographs on social media. The project's Whistle App, when completed, will allow those who witness and record human rights violations to have their information confirmed and securely forwarded to the appropriate NGOs. It will do this with a "civilian witness interface" that prompts users to supplement their material with metadata such as place, time, and other supporting information. The program will then corroborate submissions sent through the app by cross-referencing the information against third-party sources and tools, the most basic being weather and geolocation databases. The Whistle will mediate in this way between witnesses, who might not know how to produce verifiable reports, and NGOs that lack the resources to verify them independently.[16]

Other initiatives in support of uncensored internet access have been implemented by technologically cutting-edge NGOs (or programs within NGOs), of which the following are a few noteworthy examples:

- Reporters Without Borders has initiated a project titled Operation Collateral Freedom that replicates or "mirrors" websites that have been dismantled by information-censoring states.[17]

- Front Line Defenders was honored with a 2018 United Nations Human Rights Prize for its work in providing support (via an emergency hotline) for the security and protection of human rights defenders at risk, including work in building physical and digital security and strategic communications, international legal advocacy, and campaigning.[18]

- GreatFire monitors and challenges internet censorship in China through the use of circumvention tools, restoring and republishing censored information on WeChat (China's ubiquitous and popular social media application), setting up an uncensored version of the state-sanctioned internet platform Weibo (under the title FreeWeibo) with censored material restored, and distributing government-censored books by "disappeared" publishers as e-books and audio books.[19]

Taken together, these kinds of initiatives show that new technologies are acting to shift dissent and campaigns of justice toward a greater reliance on specialized technical knowledge, institutional infrastructural support, and donor financing. And this, in turn, means that activism and identity are more and more shaped by the specialized knowledge of technologically empowered elites.

Crowdsourcing Evidence

There is one sense, however, in which the elitism of human rights fact-finding takes the form more of collaboration with those who are witnesses of mass crimes, above all by drawing from source material that they directly produce. The global distribution of mobile devices, together with the vastly increased data storage capacity in the cloud, have combined to bring about a revolution of sorts in the evidentiary foundations of human rights claims. This takes the form of a vastly expanded archive of data, one of the fundamental building blocks of the world's digital architecture. The world's computing infrastructure is based on two complementary software objects—data and algorithms—that combine to produce the digital information that is consumed and put to use.[20] While much of the world's attention is going to the algorithms that form the foundation of new artificial intelligence tools, these are merely part of the final sequence of operations that a computer executes to accomplish a given task. The material on which such complex algorithmic operations are based is the data structure: the information that is digitalized in a particular way to facilitate search and retrieval.

The fact that much of the world's population now possesses mobile phones equipped with digital cameras and recording devices connected to an internet architecture supported by cloud computing has changed the nature of data accessibility and, hence, human rights fact-finding. For one thing, as-it-is-happening photographic and video evidence can be significant in proving the occurrence of and responsibility for acts of mass violence. This civilian-witness evidence can then move through the chain of human rights advocacy to journalists, NGOs, policy makers, and publics, sometimes through digital crowdsourcing projects or reporting applications.[21]

The applications developed in support of citizen witnessing and crowd-sourcing constitute a distinct genre in the emerging digital forensics toolkit. Cambridge University's The Whistle, which I just mentioned, is but one of the applications situated between events and the courtroom with the intention of protecting witnesses and verifying the authenticity of their evidence. The eyeWit-ness app similarly focuses on verification and preservation, with information recorded through the app labeled with metadata (such as the time and place of the recording), uploaded to a secure server, and then subjected to expert analy-sis.[22] There are others besides, each with a distinct solution to the problems of security, storage, and verification.[23]

It is important to beware of celebrating the development of such applications, however, before considering the extent to which they are actually distributed and put to use. The typical development of these apps involves participation of select users, without the kind of mass distribution that would reach "ordinary" witnesses. They tend not to reach those who are the targets of repression and violence, whose witnessing occurs in conditions of insecurity outside professional communities of journalists and activists. To victims, the cell phone and social media platforms remain unadorned tools of witnessing, without the protections and contextual verifications designed by activist engineers.

This is not to downplay the significance of participatory evidence gathering. Even where it has operated in the absence of tools of encryption, verification, and secure storage, it has shifted the flow of expertise. Large NGOs like Human Rights Watch and Amnesty International once typically engaged in human rights fact-finding as a process that happens "somewhere else" to people who serve as witnesses through their testimony—the raw material with which professionals then build a case. When these same witnesses are technologically empowered, however, the pattern of investigation changes. Where once there was a clear difference in authority (the "expert" conducted interviews of the "subjects" and built the dossiers of human rights fact-finding), there is now an opportunity for the subjects of human rights violations to actively participate in the inves-tigations that concern them. The information these witnesses provide is now more immediate and detailed than victim narratives drawn from memory. These advantages are creating conditions for a phenomenon variously referred to as "citizen witnessing" or "participatory fact-finding," a form of bottom-up claiming of rights, with human rights investigation transformed into a tool for community mobilization and what some see in democratic terms as "constituency building."[24]

Some of this citizen participation involves tech-savvy enthusiasts—not experts per se because they might not have specialized education, but people who dedicate time and energy to online assemblage of information. There is a vast number of amateur enthusiasts who draw from, collate, and post information online, people who, for example, dedicate days upon days of screen time to follow the movements of navy vessels. (Yes, their movements are secret, but the movements of the tugboats that bring the warships into harbor are not and can be followed by people with the knowledge and patience to access this information online. Others visit harbors and post images of navy vessels as they come into port.) Oil tankers can be more openly followed using a combination of transponders and satellite imagery (they are large enough to track using low-resolution but regularly updated satellite imagery); amateurs using TankerTrackers.com sometimes delight in their discovery when a tanker suddenly turns its transponder off, changes course, and heads to a new destination in violation of embargoes. Flight trackers are a similar resource with their own base of amateur enthusiasts. When Donald Trump and First Lady Melania Trump flew to Iraq to pay a surprise visit to the troops on 26 December 2018, the false call sign used to hide the identity of Air Force One was not enough to prevent the rapid identification of the mystery aircraft and instant global dissemination of information about where the plane was headed, so that when it landed in Iraq the event was already publicly known.

There is a segment of humanity that takes inordinate pleasure in uncovering and acting on knowledge like this, especially when it exposes a state secret or an injustice. And this, in turn, is a foundation for the emerging phenomenon of open source intelligence (OSINT), to which I turn next.

What Is Bellingcat?

One of the surefire ways to know that you're onto something new in the digital advocacy world is when the people closest to a new phenomenon can't quite identify what it is that they've created. This is certainly true in the case of Bellingcat, an organization dedicated to investigation through the tools of open source intelligence (OSINT). (The name comes from the fable of mice challenged to put a bell on a troublesome cat, with the upside-down question mark replacing the letter *i* in its logo, to suggest its dedication to difficult inquiry). The byline of the Collective (as its members also refer to it) is "Truth in a Post-Truth World." A more complete, though less catchy description would be "truth using open source tools and data." But its data is not always open source because its researchers sometimes use technologies that are "closed," accessible only to law enforcement, such as the CCTV images provided by a former police investigator who still had access that Bellingcat used in the investigation of the Skripal poisonings by Russian assassins (combined with a comprehensive database of Russian passport data, downloaded using the file-sharing platform BitTorrent). In this, shall we say, more or less "open" approach, Bellingcat explicitly sets itself apart from state intelligence agencies that develop and make use of "closed" tools oriented toward uncovering secret information. It also distinguishes itself from communities of hackers like Anonymous that are driven by the challenge breaking through firewalls and accessing secrets. Bellingcat researchers take pride in asserting that they do not assume false identities or enter into private or encrypted sources.

Then again, neither is Bellingcat oriented exclusively toward "intelligence"; it also does things that are widely recognized as advocacy, such as its workshops that train others in the techniques of open source investigation. But it doesn't stop there. The hybridity of the organization comes out when one considers that advocacy doesn't at all describe the evidence it provided the International Criminal Court (ICC) that led to the indictment of the Libyan Commander of the Al-Saiqa Brigade, Mahmoud Al-Werfalli, on charges of the war crime of murder, the result of an effort that clearly looked more like a police investigation than anything else. This investigation (which I discuss more fully below) resulted in the first ICC indictment based largely on video data originating from social media.

Building on this toehold in international law, Bellingcat's founder Eliot Higgins has begun to develop a partnership with the Human Rights Center at the University of California, Berkeley's School of Law. The group is also collaborating with the Global Legal Action Network (GLAN) in putting together the evidentiary foundations of legal claims, with a focus on the war in Yemen. And a soon-to-be-opened office in The Hague will help cement ties with the ICC, which is located there, even though Bellingcat maintains a strict policy of independence from any state or international agency or organization. The ICC's Technology Advisory Board has taken an interest in making wider use of open source tools, and Bellingcat has done presentations to inform prosecutors and judges on how OSINT can contribute to their work. (Judges in particular tend to be unaware of its potential and sometimes have to learn the basics about, for example, how Google Earth works and what makes it reliable as a source of evidence.)

An archiving platform is also in the works, based on the idea that the internet is constantly in flux, with material coming and going, but that "twenty years from now we'll have copies of vast amounts of material in searchable archives for other investigations and accountability work."[25] Higgins has at the same time begun the process of hiring staff, including the initial appointment of a staff member to be based in Colombia, which will expand the Collective's reach into Latin America.[26] With all these new initiatives happening at once, there is a sense, as Higgins put it, that "we're more or less building the plane as we're flying it."[27]

Might Bellingcat then be a kind of incipient social justice organization? But even the more encompassing term *social justice* doesn't quite cover it since the Collective has also taken on the investigation of individual crimes that are not really "social"—unless of course you consider the murder of journalists by state secret services an issue of common concern—in which case, yes.

Nobody likes to live in a condition of indefiniteness, and Bellingcat members are continually asked by the media to say a few brief words about what their organization does and where it comes from. In response to these pressures, there is an official story. Like all good origin stories, this one has a central character, a founding figure, Eliot Higgins, who began his career as an investigator in 2011, when the funding for work he was doing for refugees and asylum seekers dried up and he had extra time on his hands. His curiosity took him to the fighting in Libya during the fall of the Gaddafi regime, where he soon discovered that the tools of geolocation (he didn't know the term at the time) were useful in pinpointing events that journalists on the ground readily overlooked. (He could, for example, see the destruction of the town of Tawarga as journalists were driving past it, before the rising smoke caught their attention.) Working his research efforts into a blog in 2012, operating under the pseudonym Brown Moses (taken from a Frank Zappa song), he compiled hundreds of short video clips of the Syrian war from the internet, localized them, and analyzed the weapons used. Without speaking a word of Arabic and with no training in journalism or political science, he was able to become, in his own words, a "one-man intelligence agency," working out of his home and a sparsely furnished office in Leicester, equipped only with a laptop.[28] First, from publicly available online information he was able to compile convincing evidence that the Syrian government had used banned cluster munitions and chemical weapons. Then in 2013 he followed up by linking the Ghouta chemical attack to Syrian president Bashar al-Assad.

Higgins founded the Bellingcat website in July 2014, supported with funding from Kickstarter, with some 1,700 contributors providing £50,000 (or $70,000). With this funding and a new public identity, Higgins and his staff of four researchers pulled off a series of sensational high-profile investigations. One conclusively proved Russian responsibility for the downing of Malaysian Airlines Flight 17 over the eastern Ukraine (widely known as the MH17 disaster), in which 298 people were killed; in the process, they exposed Russian agents' awkward attempts to alter photographic evidence of the event using Photoshop. The commercial airliner was shot down on July 17, 2014, just three days after the Bellingcat collective came into being. "I was actually doing the ironing when the story broke on TV," Higgins recalled. "It was one of those moments when the world shakes slightly on its axis, and what happens next is anyone's guess."[29] Since the majority of those who died in the attack were Dutch citizens, Bellingcat's fame in The Netherlands grew in proportion to the saturation media coverage of the event. This eventually resulted in the addition of Dutch research staff and current plans for opening a branch office in The Hague.

The poisoning of former Russian military officer Sergei Skripal and his daughter Yulia in Salisbury, England, using the nerve agent Novitchok was the Collective's benchmark investigation in the UK. Bellingcat conclusively identified the Russian agents involved in the poisoning in reports sardonically titled "Anatoliy Chepiga Is a Hero of Russia: The Writing Is on the Wall"[30] and "Skripal Poisoning Suspect Dr. Alexander Mishkin, Hero of Russia."[31] Efforts to cover up the responsibility of Russian agents by having them pose as tourists in staged interviews were undone by a step-by-step log of their movements in London, accompanied by records revealing their identities as high-ranking Russian agents. (Bellingcat, in collaboration with the BBC, later identified a third suspect in the case, a high-ranking GRU operative, Denis Sergeev, in a follow-up investigation that systematically tracked his movements in London. The main source of new information came from telephone metadata records obtained from a whistleblower inside a Russian mobile service provider.)[32]

There are also unofficial stories about Bellingcat and its founder that are, in their own way, just as revealing as the narrative that members of the Collective are comfortable telling. In a clear indication that Bellingcat's inquiries had uncovered important truths, the Russian news agency RT (Russia Today) began a smear campaign against Higgins, using every possible effort to discredit his work, such as highlighting his "lucrative career as a payments officer at a women's underwear company" at the time when he worked as a blogger—everything except addressing the evidence on its own terms, which has been quietly left to stand on its own. Higgins responded, perhaps inadvisably, with a Twitter post (since removed) telling *RT* to "suck my big Bellingcat balls," a post that, in turn, formed the basis of yet another line of attack in the *RT* campaign.[33] Switching to the language of a Bellingcat report, Higgins offered a more effective rebuttal: "Both Russia and Syria have a dubious reputation for factual reporting on the issue of chemical weapons due to their accusations of vast international conspiracies, use of doctored satellite images, and tendency to present videos and images from computer games as evidence."[34] When I asked Higgins about this experience, he speculated that *RT* probably saw Bellingcat as just another blog, overlooking its connections to journalists who were already supporters and for whom the smear campaign came across as more evidence that the investigations had hit their mark: "All it did was reinforce the idea that we were on to something."[35]

What is clear from any angle one looks at it is the continuing success of the Collective and its community of (in large measure) former gamers and programmers dedicated to the challenge and truth-value of open source investigations. In recognition of his high-profile investigative work, Higgins was awarded a 2018 Magnitsky Human Rights Award (named after the Russian anticorruption accountant, Sergei Magnitsky, who died in 2009 after spending eleven months in police custody). More important than the accolades, he gained the kind of exposure that brought in donations and allowed the organization to grow. With the Collective's growing fame came grants provided by Google, the Open Society Foundation, Meedan (a nonprofit oriented towards cross-language interaction on the web), and the Adessium Foundation, among others. Bellingcat researcher Pieter van Huis explained to me that the Collective has also benefitted financially from a decision made a few years ago to charge for its booked-months-in-advance workshops rather than doing them for free. Participants are asked to apply, as I did, with an emailed explanation of their purpose in taking the workshop, although the only strict stipulation is that participants must not be working for any secret service organization, which would tend to have an inhibiting effect on others who might want to participate. At $2,400 per person and with between fifteen and twenty people regularly filling the rosters of workshops held in London, Amsterdam, Toronto, Munich, Zurich, Toronto, Washington, Berkeley and San Francisco, and, more recently, Lima, Medellín, Bogotá, and Mexico City, the Collective is now doing, as the British might say, quite nicely, thank you.

So much for how it got started; but what does it actually *do*? Its goal is formally "open source analysis, with transparency with our sources and how we reached our conclusions."[36] Or, expressing the same idea even more formally (ironically, in a Twitter post), it aims "to create a replicable methodology and a comprehensive public database to meet the highest evidentiary standards for potential use in impact-oriented investigative journalism, academia, and legal endeavors."[37] What this means in practice is close attention to controversial events using evidence from online sources, addressing the question concerning *what* happened and not *why* it happened. Human sources are considered fallible, though journalists are certainly welcome to supplement Bellingcat's work with their own field inquiries using the usual methods of observations and interviews. Often, however, circumstances of state repression prevent journalists from traveling to sensitive areas, and this is where Bellingcat's work really comes into its own. It can make creative use of readily accessible technologies to conduct investigations that would otherwise be impossible, into events that would be subjected to the confusion of the moment or the dampening effects of state propaganda.

The principle of transparency and the solidity of the evidence it amasses in its investigations became evident to me from the exercises assigned to participants in a workshop I attended in Amsterdam in February 2019. One of these consisted of geospatial analysis of a photograph, a view from inside an airplane of the wing and the landscape it was flying over. With nothing more than the logo of Ukrainian International Airlines on the winglets and a view of the ground to go on, our task involved geolocation using Google Earth Pro to identify exactly where the plane was flying (just south of Kiev, the participants soon discovered). A related exercise in geolocation used Google Maps in its regular search mode and in Street View to find telltale indicators like a barely visible storefront and, from these obscure clues, to identify the setting in photographs posted on social media by self-declared ISIS members in Paris, depicting notes affirming their fealty to the organization. (This exercise was based on an actual event in which French police were conveyed information about the specific apartment from which the photograph was taken.) Sticker graffiti on a pole, visible in one version of Street View and gone in another, or the price of gasoline on a sign, can establish the approximate time an image was taken.

The workshop got progressively more complicated, to the point that Henk van Ess, one of the foremost experts in social media investigation, took participants step by step through the methods he uses to uncover private phone numbers, in this case leading to the discovery of what was likely the private cell number of Michelle Obama. (We were advised not to use or make note of it.) He then instructed participants how to splice in and edit lines of code to link searches between Facebook and Instagram. By the end of the workshop it was abundantly clear that the tools used in open source investigations are, under the right circumstances, able to provide strong evidence of mass crimes, even in circumstances in which investigators are unable to travel to the location in which they occurred.

The steps taken in gathering open source evidence can be reproduced in a scientific way, without requiring access to a laboratory or expensive field research. As van Huis put it, "You have to show things in such a way that the public can follow you. In the end you have to explain it to someone on the other side of the world, and you have to explain it quickly." Consistent with this advice, Bellingcat's online reports can be followed by anyone with a modicum of computer skills and the patience to follow the evidence trails step by step. This is the basis of the new form of journalism that Higgins has characterized as "a peaceful revolution taking place without weapons, but with smartphones."[38]

This new approach to investigation has produced a torrent of research output, which is gaining in momentum as the Collective expands. From the launch of its website in 2014 to the present, the site has produced to date approximately 750 posts or reports on investigations—the journalistic equivalents of a "story" but with explanatory notes that make it reproducible in the manner of a scientific article. Approximately 350 of these have been translated and posted on Bellingcat's parallel Russian website, ru.bellingcat.com.

At the origin of this research, it has a formal structure that is key to its success as an organization, with a four-member management team headed by Eliot Higgins as chairman/executive director, supported by a business director, head of training and research, and a project manager. A staff of eighteen full-time employees (up from four when the Collective began) is supported by a global network of some thirty regular contributors. In a move toward greater formalization, its core members drafted a policy plan and registered the organization as an NGO with charity status in Amsterdam in July 2018, under the Dutch name Stichting Bellingcat, or Bellingcat Foundation.[39] From an informal start in the blogosphere (about which its members take great pride), it very quickly took on the trappings of an ambitious, well-organized international group.

But even with its formal organization established, the boundary between Bellingcat and not-Bellingcat isn't that clear. In an unofficial direction, the Collective's core members often have professional networks that remain as resources, such as those hired by major media outlets who remain as well-connected allies.[40] Christiaan Triebert, senior investigator and lead trainer at Bellingcat, pointed to this close connection with mainstream media: "Either journalists [request] our work or we send it to them. They convey what we do to a large readership in a way that we can't." (Consistent with what he said, Triebert offered this insight shortly before taking up a position as an investigative journalist with the *New York Times*.) Staff members and researchers with histories of employment in law enforcement provide another area of contribution. Bellingcat's collaborative relationship with law enforcement is helped along by its regular assistance to Europol (the European Union Agency for Law Enforcement Cooperation) in its investigations into the sexual exploitation of children.[41]

The Bellingcat Collective is by no means the only open source collective but collaborates with other organizations that use (and in some cases develop) tools of open source investigation. Each has its own area of specialization. One of these, Syrian Archive, composed mainly of Syrian refugees and human rights activists in Germany, is dedicated to "curating visual documentation relating to human rights violations and other crimes committed by all sides during the conflict in Syria with the goal of creating an evidence-based tool for reporting, advocacy, and accountability."[42] It has worked with Bellingcat researchers in an ambitious documentation project that to date has collected over 3,300,000 items of digital content, of which it has verified and archived 5,743 in evidentiary support of claims that cover nearly the entire gamut of gross human rights violations and war crimes, including use of illegal weapons (e.g., chemical weapons and cluster bombs), massacres and other unlawful killings, sieges and violations of economic, social, and cultural rights, hostage taking, arbitrary and forcible displacement, violations of children's rights, torture and ill treatment of detainees, and sexual and gender-based violence.

Another investigative organization, the Atlantic Council's Digital Forensic Research Lab (DFRLab), is oriented more toward preventive efforts, by "operationalizing" the study of and intervention in digital disinformation, "exposing falsehoods and fake news, documenting human rights abuses, and building digital resilience worldwide."[43] It approaches its tasks in part through crowdsourced investigations conducted under #DigitalSherlocks.[44] Although it overlaps with Bellingcat in its concern with, for example, the siege of Aleppo and ceasefire violations in the Ukraine, it has also, unlike Bellingcat, assigned itself the mission of promoting "digital resilience" by tracking disinformation campaigns and efforts to thwart or interfere in democratic process, as well as going to the root of these problems with education programs, teaching the "public skills to identify and expose attempts to pollute the information space."[45]

Finally, the network of expertise in which Bellingcat is situated includes a small number of architectural firms that have developed sophisticated forensic techniques of geospatial analysis for use in the courtroom. Two of these organizations, SITU Research and Forensic Architecture, specialize in analyzing and formally presenting OSINT evidence. Having developed this expertise with an expanded range of research products, they have worked with investigators and prosecutors in a number of high-profile cases, including at the International Criminal Court.

Based in a warehouse in the Brooklyn Navy Yard, SITU Research's staff profiles include architects, urban planners, tech experts, designers, policy experts, and a dog, Lola, listed as the director of office culture. SITU's clients include Amnesty International, Human Rights Watch, the Bureau of Investigative Journalism, the Office of the High Commissioner for Human Rights, and an international scattering of law schools.[46] Brad Samuels, founding partner of SITU, described to me the insight that he and three colleagues had upon their graduation from the Cooper Union that led to SITU's founding in 2005: They could expand the scope of their work as architects by doing spatial analysis of past events in addition to the usual future-oriented prospective design of urban space.[47]

This was the moment of inspiration that eventually led to SITU's participation (together with Forensic Architecture) in the 2009 investigation and trial following the death of thirty-year-old Bassem Abu-Rahma, a Palestinian man killed by a tear gas canister during a peaceful protest in the West Bank village of Bil'In. This marked the first time that one of SITU's research products was presented in court (in this case, an Israeli military tribunal). At the request of attorney Michael Sfard and the human rights organization B'Tselem, SITU created a reconstruction based on civilian video footage, showing conclusively that the tear gas canister that struck Abu-Rahma in the chest was fired at close quarters directly at the protester (rather than at the 60-degree minimum angle mandated by the Open Fire Regulations for indirect fire).[48] For the purposes of the military tribunal, SITU and Forensic Architecture developed a "parametric tool" that "allowed key variables in the ballistic equation to be easily modified, updated and visualized," offering conclusive evidence that the gas canister could not have been fired at an angle of 60 degrees or more.[49] Despite this evidence, the case was summarily closed by Major General Danni Efroni after a delay of three years after the trial.

The team of investigators at Forensic Architecture, an independent research agency based in Goldsmiths, University of London, is, if anything, even more eclectic than that of SITU Research, made up of architects, artists, filmmakers, software developers, journalists, archaeologists, and lawyers.[50] An example of its work can be seen a more recent case involving video footage of a Palestinian killed by Israeli forces, the subject of a short *New York Times* documentary on the killing of medic Rouzan al-Najjar by an Israeli sniper while she was attending to the wounded during the 2018 protests against Israel's blockade of the Gaza Strip. The event offered ideal conditions for an open source investigation, with a controversy (Palestinian versus Israeli narratives of the event) following from a specific violent incident in the context of a media-saturated environment. Over a thousand photographs and videos were assembled and analyzed over a period of more than six months. Forensic Architecture then provided the capstone of the investigation: a 3-D model of the fatal moment, in which the bullet ricochets off the ground and strikes Rouzan in the chest. It shows conclusively that, though she may not have been directly targeted by the Israeli sniper, none of the medics in the line of fire posed a threat to the Israelis and that the shooting that resulted in Rouzan's death "appears to have been reckless at best, and possibly a war crime, for which no one has yet been punished."[51]

The difficulty of putting one's finger on the essence of Bellingcat is not just a good topic of conversation for members of the Collective over beers but goes to the heart of what constitutes the contribution of technologically empowered individuals and organizations in the emerging era of human rights. Posing the question of Bellingcat's relation to human rights brings us into the same territory that we found ourselves with the inquiry into graffiti in chapter 1. Clearly, encouraging human rights compliance through OSINT and public access to its results constitutes only part of what it is and does; yet understanding what Bellingcat is and does brings us to the heart of how human rights work in practice. In particular, it brings more clearly into focus the powerful motivation that comes not directly from human rights law but from a sense of justice relative to what it is to be human with dignity. This is something that puts people in harm's way in war zones to witness and record bombing patterns and that puts people for days at a time behind computer screens, where they obsess over things like piecing together crowdsourced video of an incident of police brutality or digitally sharpening, frame by frame over a period of weeks, a blurred CCTV image of the license plate of suspected murderers of a journalist—to the point of burnout.

How to Do a Bellingcat Investigation

Bellingcat workshops are another way the Collective is expanding its network. Every workshop brings together fifteen to twenty people from the fields of journalism, corporate security, law enforcement, diplomacy, NGO activism, and, in the workshop that I attended in Amsterdam, someone looking for open source techniques to help solve the murder of a close family member. At the same time, Bellingcat publishes widely available guides on how to conduct investigations and how to publish reports. Its titles include *How to Scrape Interactive Geospatial Data,*[52] *How to Identify Burnt Villages by Satellite Imagery—Case Studies from California, Nigeria, and Myanmar,*[53] and *Creating an Android Open Source Research Device on Your PC.*[54] These serve as instructional reference points for anyone who wants to learn OSINT investigation and reporting methods, including those who want a refresher on the instruction they received in a workshop. A good portion of its (to date) 750 investigative articles have been done by former workshop participants, posted on the website after being vetted by research staff and, usually, revised in response to their review. With each workshop it offers, the number of people familiar with its tools and techniques grows and the Collective gains in reach, influence, and output.

I signed up for the Bellingcat workshop in Amsterdam already knowing that I wanted to do my own investigation, preferably something to do with the crisis in the central Sahara and the fact that the UN peacekeeping mission in Mali is the most dangerous in the world. (I discuss this conflict further in chapter 5) The ideal Bellingcat investigation, as I mentioned above, involves a specific event that produces a controversy, with different versions of the event in conflict, and with an abundance of photo and video evidence available online. There were plenty of specific incidents for me to choose from, many of them involving coordinated attacks on UN compounds, but there was little I could find in the way of a question to be addressed out of the few still-posted images of wreckage, mangled corpses, and blood-soaked sand. Most of the evidence from such events gets taken down from YouTube and other social media platforms within days or hours of getting posted. (YouTube, for example, has a policy against users posting content that depicts "road accidents, natural disasters, war aftermath, terrorist attack aftermath, street fights, physical attacks, sexual assaults, immolation, torture, corpses, protests or riots."[55])

Given these limitations, I eventually settled on the topic of child soldiers among the Tuareg insurgents engaged in violent conflict in northern Mali since

2017, when UN sanctions were first applied to organizations and individual leaders responsible for gross human rights violations in the region. This is not, from the Bellingcat perspective, an ideal topic since it isn't tied to a specific incident, but it did meet the criterion of being based on a controversy: a UN report that attributed specific numbers of child soldiers to various insurgent groups and public denials from spokespeople of the organizations that they are using child soldiers. I was assisted in my investigation by four workshop participants: a head of security for a transnational corporation, two recent graduates of a master's program in journalism, and Bellingcat's newest staff member, Rawan Shaif, who self-identifies on Twitter as an "investigator and peripatetic journalist." (Her self-description on Twitter continues, "Come for Weapons, War, Armed Groups and all things Yemen related/adjacent. Humor OK.").

Shaif's appointment to the Bellingcat staff represents another direction in the expansion of the Collective's range of interests, in this case focusing on NATO-supported bombing campaigns in Yemen. As the staff member assigned to this important dossier, her peripatetic days were over, or at least much reduced, and she was recovering from exhaustion, having just finished organizing a four-day Yemen-oriented hackathon hosted by University College London, "an unbelievable way to kick off my first week working at @bellingcat."[56] She was amazed at what the hackathon had accomplished using open source tools and techniques.[57] With nothing more than a close analysis of satellite imagery, for example, participants were able to identify closely grouped "double strikes" that people on the ground saw only as one massive explosion.

I didn't know anything about Shaif before she was assigned to my table to help with the investigation of child soldiers in the central Sahara. (My topic had been accepted as one of three to be taken up, the other two involving a murder inquiry and an incident of police violence during the yellow jacket protests in Paris.) I was impressed by the way she quickly organized the group, explained the steps to be taken in conducting an investigation, and, within fifteen minutes of searching, found what appeared to be a recruitment video featuring Tuareg youth. I conveyed my favorable impression to Christiaan Triebert and he agreed: "She visited secret joint forces raids on the ground in Yemen that nobody knew about. She's a badass. Figuring out raids that are not publicly known, now that takes courage."

The first step involved in this kind of investigation is to find and download everything related to it online. In events like a chemical bombing attack, the most graphic imagery, which is at the same time the most useful, will be removed from the platforms almost right away, so it is a race against time. Tools like KeepVid. com for use with YouTube and FBdown.net for Facebook are the basics, which can be supplemented by other streaming-video recorder software and scripts such as youtube-dl, a command-line program used to control other video downloading software applications. The downloaded material has to be meticulously organized and documented with URLs so that, if any are later found to be useful, they can be added to the report to make the steps taken in the research process transparent, even if the material is no longer available online.

The search for material (photographic and video) usually involves systematic navigation of social media. In the case of the child soldier investigation, this involved starting with the leaders of the organizations identified in the UN report and expanding outward, finding connections between people from their Friends lists, looking at their posts and—not to be forgotten—the discussion threads that accompany relevant material. One video I discovered showed children who looked to be between twelve and fourteen years old on military parade, but it was set in a featureless desert landscape, which would have been impossible to geolocate using the usual methods. However, in one of the last thread comments, a contributor remarked simply, "It is in Kidal!" and no other information was needed. I noticed that the farther one got from the leadership, the less the material seemed carefully curated. Shaif made this observation more pointedly: "It's always the bottom feeders that give you the best information."

This gave me a head start (at least with that one video) into the next stage of the investigation: verification. For the purposes of the workshop, the process of verification could not go far, and to illustrate this part of the investigation I will leave to one side my focus on the central Sahara and turn to another example: the war crimes investigation of events in Libya submitted to the International Criminal Court. It sometimes takes several years to amass answers to the questions associated with video or photographic imagery. These include its origin, who is the source? Where was the video or photograph taken? When? And why? Was there a political bias behind the posting of this material? Behind these questions is a constant, nagging suspicion that material might have been faked. Journalists and bloggers are lazy, and people often take old footage to represent a current event. Some videos pop up online again and again. Authors of controversial material will sometimes try to cover their tracks using, for example, GPS spoofers that send false data to their GPS service to hide their location. In their approach to material from social media, lawyers and judges are acutely aware of these possibilities, so establishing the verisimilitude of material is essential, particularly concerning when and where the photo or footage was taken.

Lawyers and judges often have understandable difficulty accepting new methods and sources of knowledge, and it has taken efforts from organizations outside the court to make room for OSINT in international criminal and human rights investigations. Some of the difficulties of translating open source evidence for these investigations, for example, are being addressed through an initiative in the Human Rights Center at UC Berkeley Law School aimed at the development of protocols that will "set common standards and guidelines for the identification, collection, preservation, verification and analysis of online open source information."[58] These protocols would guide NGO representatives and first responders in the collection of evidence in ways that would make it consistent with legal requirements and admissible in court. It would also help lawyers and judges better understand how to evaluate evidence based on open source investigative techniques. Finally the protocols are intended as a tool for investigators, probably less useful in their areas of strength—developing investigative leads and gathering evidence—but intended to develop those aspects of investigations in which they might have less experience, that have less to do with the immediate rush and gratification of the chase, things like assessing security risks and protecting witnesses.

Even in the absence of these protocols, the International Criminal Court has made use of OSINT in several prominent investigations, assisted by a variety of organizations with the requisite technical knowledge. Open source evidence as a central aspect of an ICC prosecution first took place in the trial of Ahmad Al Faqi Al Mahdi, a member of the Tuareg Salafist group Ansar Dine (which I discuss at greater length in chapter 5), who was charged with participating in the destruction of cultural heritage (nine mausoleums and a mosque door) in Timbuktu during the Malian civil war in 2012. Because of the court's lack of experience with such evidence and the absence of protocols guiding its use, the task of assembling the material was given to SITU Research (introduced above).

The platform developed by SITU in close collaboration with the ICC's Office of the Prosecutor allows users to select individual sites from a satellite image of Timbuktu, with all the associated material accessible either simultaneously or as individual articles.[59] Its design was intended specifically for use in the courtroom. Though it was put to use in a case involving destruction of cultural heritage—an ideal use of SITU Research's work in the intersection of architecture, urban planning, policy, and human rights—the Al Mahdi prosecution represents an early step in a wider movement toward use of digital technologies in international judicial proceedings, particularly those aimed at bringing accountability for crimes of mass atrocity. Al-Mahdi, whose trial began on 22 August 2016, quietly pleaded guilty to the charges of destroying nine mausoleums and a mosque and made a statement in which he expressed remorse and advised others not to follow his example. He was sentenced to nine years in prison.

The evidence compiled in the investigation of Mahmoud Mustafa Busayf Al-Werfalli, Axes Commander of the Al-Saiqa Brigade in Libya, for his leadership role in mass executions of terror suspects involves an unremorseful, repeat offender and offers an even clearer example of how the process of verification of open source evidence can be done. More broadly, this case confirmed the significance of social media material as the possible basis for ICC indictments. The court's justification for pursuing the indictment for the war crime of murder in the Al-Werfalli case stemmed in part from "the posting on social media of the videos depicting the executions, and the frequency and particular cruelty with which they are carried out."[60] In the course of a Libyan antiterrorist campaign titled Operation Dignity, launched on 16 May 2014, Al-Werfalli (clearly identifiable in videos posted online, wearing a black cap, camouflage pants, and a black T-shirt with the logo of the Al-Saiqa Brigade) is depicted personally killing prisoners wearing orange jumpsuits with their hands tied behind their backs. He is also seen ordering their systematic execution by men under his command. Video material posted by the Media Center of the Al-Saiqa Brigade and transcripts of this material were key to the investigation. In other words, the investigation was based on a set of strange and disturbing acts of self-incrimination that an astute Twitter user (retweeting @BBCArabic coverage of the story) characterized as "war crimes for likes."[61] One mass execution in particular, "incident seven" in the indictment, offered convincing evidence of Al-Werfalli's war crimes. The video of this incident, involving a total of twenty executed persons, was posted on social media on 23 July 2017. Bellingcat took on the challenge of authenticating this video and converting it into admissible evidence.[62]

The first thing to establish was the location of the executions. Images depicting landmarks and buildings in the surrounding area were visible in satellite imagery from Google Earth Pro, which then pinpointed the area and provided a geocode. The images were combined using the software tool Agisoft Metashape to get a wider field of view and determine the exact camera location. This information was then traced to a military compound under the command of Al-Werfalli. All this information answered the questions about the location, but it did not deal with the possible defense that the images were faked, that the whole thing was staged. To address this doubt, satellite imagery was again put to use, this time using the archive of images that provide a kind of "time machine" record of the surface of the planet. Here is where the investigators got lucky. TerraServer imagery from the period just after the executions took place revealed a pattern of dark markings that exactly matched the position of the prisoners in the execution video—presumably bloodstains. This finding not only linked the satellite imagery conclusively to the video but gave an approximate date when the executions took place, between 14 and 17 July 2017. This connection between satellite imagery and the video material downloaded from YouTube was key to the decision by prosecutors to accept open source material as a foundation for the indictment. As the ICC expressed it in its inimitable way, "The Chamber is satisfied that the above mentioned video has sufficient indicia of authenticity in order to be relied upon at this stage of the proceedings."[63]

Al-Werfalli was arrested soon after the indictment was issued but then "escaped" from prison through the apparent collusion of the Libyan army, even though the evidence against him is clearly enough for a conviction.[64] Soon thereafter, he was again seen online committing more executions.

Powers, Limitations, and Imaginaries

It is tempting to imagine hopefully that the transparency of Bellingcat's methods can offer a way through the confusions of post-truth media, but things are not so simple. The Russian government, which once derided Bellingcat's methods while having its skullduggery exposed by the group, has begun to issue its own satellite imagery in support of its version of the truth, leaving it to the public to separate genuine from faked material. And crowdsourced Reddit investigations of the kind that produced the famous conspiracy theories associated with the terms *Pizzagate* and *QAnon* make use of "evidence" from Google Maps or Facebook presented in ways analogous to a Bellingcat report.[65] The rise of troll farms specializing in "hacked materials" and disinformation in the post-truth media ecosystem means that Bellingcat's impact depends on what might be called *open source literacy*, the public cultivation of techniques for reading a detailed report or taking apart spurious evidence using open source tools. The difficulty is that checking sources and repeating an investigation are hardly the stuff of mass appeal. Most people will ultimately either take Bellingcat's investigations at their word or dismiss them according to their beliefs and political loyalties.

Yet there is in all of this an emerging picture of a new relationship between technological expertise and the pursuit of justice. Open source intelligence is growing exponentially through public channels, with "citizen journalists" acting on their senses of curiosity, technological empowerment, and justice. This has consequences for human rights fact-finding, along with investigation into every kind of crime that lends itself to the alchemy of OSINT. Its techniques are vastly more accessible than the advanced education of software engineers. As Triebert put it, "It's not rocket science, you don't need a PhD. You just have to be dedicated." And with that accessibility, a shift is taking place in the power of publics. The laptop has not quite replaced the street as a locus of dissent, but things have certainly moved in that direction. As a collective that has rapidly shifted the evidentiary foundation of public discourse, Bellingcat's work helps us witness in real time the changing nature of human rights.

These impressive capacities and the seemingly sudden, out-of-nowhere discovery of open source tools and their mass availability might easily provoke a sense that they constitute a cure-all, a solution (if not now, then in the future) to all the ills of disinformation and impunity that plague international justice. But, in the service of its ability to do what it does, we should be aware of what it cannot (and was never intended to) do.

In open source investigations, yes or no questions are commonly applied to acts of violence, interrogating things like whether a chemical weapon was used in an attack as reported by medical personnel on the ground or the denials from the Syrian government are plausible. Did a bombing take place in a Yemeni marketplace, or were official reports of a legitimate military target truthful? Such tightly focused interrogative binaries do not (because they cannot) go beyond moments of atrocity to take in the wider frame, the context of past, present, and future that helps us understand the conflict and the people involved. The moral certainty of video-captured instants of violence become the central reference point for understanding conflicts as a whole. "Browser history" is a poor substitute for more searching inquiry into the past. The flat temporality of the screen only sees power in the present but does not have a way to see how it came to occupy its place. Colonial power, as a precursor to many of the conflicts it documents, remains almost entirely unexamined.

Open source's focus on yes/no answers to specific incidents cannot readily see (and does not purport to see) the question of wider collective responsibility for the commission of war crimes and mass atrocity in the present. Because it focuses on individual incidents, there is, for example, little in the method to reveal conflicts of political values that lie at the heart of dysfunction in global justice regimes. It can sometimes expose but has difficulty counteracting the ideas and loyalties in which the crimes and disinformation it uncovers are justified and feed the demands of criminal states' true believers. It accumulates singular data points. It doesn't show the wider interrelatedness of events or the way that events are happening—particularly with reference to history—in relation to one another. Each investigation is a spotlight that only vaguely and peripherally reveals what lies beyond the edge of illumination.

The counterpoint to the absence of history is a futurism that characterizes the knowledge bases accumulated in open source investigations. The process of archiving war crimes in the hope that they might someday be prosecutable encourages forward looking, an expectation that the politics that create prosecutorial incapacity will eventually shift and those responsible for today's horrors committed in conditions of impunity will ultimately be brought to justice. The predictive algorithms in which many open source researchers are trained have an affinity with a future orientation to their work and an oddly hopeful approach to the images of horror in which they are often mired.

Of course, the remote possibility that judgment and justice will ever occur is no reason to abstain from the effort of accumulating evidence and imagining better ways in which it might someday be used. There are current benefits to be found in the future-oriented dossier-building efforts of open source investigators. Documentation of war crimes like those occurring in Syria and Yemen affirms the experiences of survivors in much the same way as the accumulations of testimony that go into truth commissions. But, more than the sympathetic response of an audience to stories of suffering, it offers *evidence* as a form of affirmation. These efforts put the lie to state-sanctioned propaganda and false reporting, giving greater clarity to the pathways of responsible journalism and public indignation. And, despite the present incapacities of international law to prosecute them, the war crimes cases that investigators are building might well contribute to pressures toward the reform of international criminal law, in which masses of digital evidence can serve not only as instruments of justice but as reference points of imagination.

CHAPTER 4

Shouting Above the Noise

Your protest should not be individual, whispered from one ear to another.
That has been done time and again, without any result. It should be
collective, made with voices raised, to everyone, on every occasion, to all
the constitutional bodies that oppress us, that exclude us, that diminish us.
Gather yourselves into associations, and let your banners carry slogans on
the two sides of the resistance, that of the workers and that of women.

Anna Maria Mozzoni, *Un passo avanti nella
cultura femminile: Tesi e progetto*[1]

Public Minds

The distinguished Italian philologist, philosopher, and historian Umberto Eco,
in a speech given on the occasion of his receipt of an honorary degree at the
University of Torino, presented a blistering critique of the intellectual climate
created by new technologies: "Social media," he declaimed, "gives legions of idiots
the right to speak, people who used to speak only at the bar after a glass of wine,
without damaging the collectivity. . . . Now they have the same right to speech
as a Nobel Prize laureate. It is the invasion of imbeciles."[2] His view, ironically,
broke through the supposed cacophony of idiocy that was the subject of his ire
and hit the newswires with global coverage, fueling widespread concerns about
the consequences of social media.

Eco was pointing to a very real, very troubling aspect of the post-truth phenomenon. The ability to bypass journalistic filters and post claim-based identities
online has gone together with the near collapse of the role of intellectuals as "leaders." The all-pervasiveness of information and the possibilities for posting one's
views, while consuming ideas according to personal inclinations, has meant that
virtually everyone in some ways acts as an "intellectual," often while refusing to be
guided or challenged in their opinions. And when *everyone* has potential access
to the world's attention, the ideas that gain currency and the popular sentiments

that put them in the spotlight are anything but refined, noble, or edifying. The worst inclinations of publics toward xenophobia, superficiality, vainglory, and voyeurism have long been there in the consumption of mass media. The problem is that now, through the power of social media, what was once an undercurrent of opinion has become more visible, laid bare by the technologies that give it global access and presence. Publics have now become the main source for the education and information of publics. The noise created by this exchange of imbecilities (to use Eco's term) drowns out the ideas of the philosophers, artists, and writers with important and edifying things to say. No one is listening to intellectuals. The peanut gallery has become the stage.

This dismal characterization of the quality of public discourse, however, is incomplete without considering some of the media-based possibilities for public outreach and persuasion and what is being done with them. Long before human rights were formalized in international law, activists noted and made use of the possibilities inherent in currents of opinion, and above all the influence that follows from shaping them. In the epigraph to this chapter, the extension of public communication to techniques of protest are captured succinctly by Anna Maria Mozzoni (1837–1920), a founding figure of the women's movement in Italy. Of course, the early struggles to capture opinion on behalf of a cause were also often met with opposition, in ways that are basically similar to many circumstances today. This has included not only the kind of outright censorship that I discussed in chapter 3 but also efforts to use the platforms of power to make a bigger noise, drown out the sounds of dissent, or to capture and coopt the "voices raised" with superficial forms of accommodation that ultimately serve dominant interests. Mozzoni did not live to see the extension of the vote to women in Italy. Her campaign for women's rights in the late nineteenth century, which was limited mainly to holding meetings and writing opinion pieces, is perhaps an early benchmark for collective justice lobbying and the power of media (and, in her case, its limits) as a way to convey knowledge of justice causes and to broadly influence public opinion.

With powerful, globe-spanning tools of communication and organization becoming available to activists in recent decades, we have seen a dramatic increase in the use of public opinion in the exercise of dissent. As the internet developed, great (and in some ways misplaced) hope was attached to the idea that communication had at last been "democratized" to the point that the means of influence over public opinion was no longer controlled quite so securely by states and other power interests but was part of widely available tools of outreach and persuasion, oriented to a transnational, rights-conscious public. Expanding the community of the committed is a strategy oriented toward empowerment, with communication to mass audiences serving to persuade publics and gather members, acting as a deterrent to political and legal abuse and an impetus toward reform. Whether one is communicating through a well-funded website or through a few simple words sprayed on a wall, the *goals* of persuasion are essentially the same.

But just who are these anonymous publics that the (often equally anonymous) authors of social justice messages are trying to reach? On the face of it, this may be an impossible question to answer. Publics are, after all, unknown almost by definition. Their members appear only occasionally in such things as rallies, flash mobs, opinion editorials, as well as in nonanonymous statements online, in Facebook posts, tweets, and (now, less frequently) blogs. The mass of those consuming information in (more or less) unison as a foundation of public discourse, however, remains significantly inscrutable, unknowable, and unpredictable.

But not everyone has been inhibited by the unknowable nature of publics to the point of refusing to speculate about their thoughts, emotions, and inclinations. Jonah Berger, writing about marketing strategy, puts the emphasis on those qualities of products that readily catch on in a way that is "word of mouth worthy." The guiding principle is simple: "Smartphones tend to be more exciting than tax returns, talking dogs are more interesting than tort reform, and Hollywood movies are cooler than toasters or blenders."[3] The entire apparatus of marketing in the promotion of consumer products is premised on finding out about the public's desires, with increasing sophistication as social media provides targeted insight into the thoughts and habits of countless individuals. Katherine Boo, in an essay written for aspiring journalists, frames the challenge of public appeal from a different angle: "When your subjects are grim and your characters destitute, disabled, or extremely unintelligent, and the wrongs against them are complicated, how many people are going to relish tucking into your story with their bagels and cream cheese on Sunday morning?"[4] In marketing, as in journalism, not every product has intrinsic appeal.

Politicians also make it their business to try to know what they can about public opinion. After all, their careers and political ambitions depend on it. Election campaigns are covered by mainstream media with detailed and seemingly desperate attempts to gauge the intentions of registered voters, with notoriously varied amounts of success. (Even the otherwise uncannily accurate FiveThirtyEight failed to predict Donald Trump's victory in the 2016 presidential race, though it had the results closer than most.)

There is also a sizeable literature that considers the influence of opinion on criminal justice, with arguments variously emphasizing the public's tendencies toward apathy and willful ignorance on justice issues or their habit of making pragmatic cognitive shortcuts.[5] This effort to understand the public influence on criminal process sometimes leads to inquiry into the essential qualities of media themselves, such as the tendency for media to commodify information, and in so doing to compromise rationality in favor of the mythical or charismatic qualities of judicial process.[6]

For reasons I can't fully explain, however, there is surprisingly little effort to gauge the significance and variety of public opinion in campaigns of human rights, environmental justice, and other forms of "soft" or "non-justiciable" law that go beyond state borders and single cases. True, the UN and Google have collaborated in the Global Pulse initiative, which in one project gauges some of the main currents of opinion on Twitter, mainly with reference to the UN's Millennium Development Goals.[7] Despite the impressive algorithms at work, however, this is a limited exploration of a complex, shifting reality using the most algorithm-accessible platform. We still know little about how (and why) human rights campaigns are received by their audiences. This is unfortunate, because acts of persuasion and collective opinion in one form or another are the essential means by which international norms have any effect at all on the behavior or states. The public consumption of information about human rights abuses and their victims and the activism that follows from success-ful persuasion are essential to the political calculus that ultimately influences rights compliance.

If the "public mind" in criminal cases is difficult to fathom, it is even more complex and contradictory in the context of international justice campaigns. Judging solely by the range of responses to news items, it is clear that as a whole this "mind" is not entirely rational. The various publics that take an interest (or active disinterest) in human rights causes tend to be emotionally reactive and at the same time jaded and skeptical. They are easily bored. They get accustomed to images of violence and suffering, even though they are voyeuristically drawn to them, like drivers who slow down at the scene of an accident. They can be critical and try to root out error in the information they receive, but at the same time they have contradictory tendencies toward simplification of complex issues and attraction toward the spectacular, the unusual, and the emotionally affecting.[8]

These are qualities of public opinion that are particularly visible in human rights processes, with claimants facing the challenge of collective self-representation to NGO activists, UN experts, and public consumers (and potential supporters) of their causes.

What conditions have to come together for new technologies to enable human rights activists to harness the powers of public awareness? The very nature of this question calls for a two-part response, since understanding new conditions of public awareness first calls for us to consider the conditions of its repression: the silencing effect of noise and the techniques and technologies used by rights-violating states, corporations, and other perpetrators in efforts to thwart public scrutiny of their actions and repress dissent. Only with an awareness of the obstacles to public awareness and solidarity can we appreciate the key qualities of successful campaigns for justice, several examples of which I turn to later in the chapter.

Ideal Victims, Ideal Perpetrators

In certain ways, justice-oriented public outreach works very much like advertising in its efforts at persuasion. Commercial marketing, however, is rarely about injustice but more often oriented toward accentuating life's possibilities—or making its wonderful impossibilities seem within reach. Just imagine how great things would be if you could have *this* car, *this* insurance plan, *this* pill. We don't see peoples' miserable lives *before* they bought that car and are never shown graphic images of suicides in advertisements for antidepressants. The persuasions of consumer society are oriented toward a piecemeal utopia, offering this and that little bit of happiness, depending on how you spend your money.

Despite this limitation, the range of emotions that can be used in commercial marketing is much greater than in social justice activism, or at least on a very different spectrum. Humor and human rights violations, for example, do not go well together. There is nothing to joke about when it comes to mass killing. Pleasurable sex is out of the question—only rape in the context of mass violence will do. Curiosity, too, can only go in particular directions. In its campaign using little-known "fun facts," for example, Snapple does not put the numbers of Somali refugees in Kenya on the bottom of its lids (527,235 as of 31 August 2018).[9] This means that justice lobbying has a limited repertoire of emotions that can be provoked to gain the sympathies and commitments of the consumers of injustice. In other words, in the context of an all-pervasive consumer society, human rights advocacy has an image problem.

It does have one tool, however, that commercial marketers don't have: the emotionally powerful sense of injustice. Furthering a collective justice claim calls for publics to feel a sense of indignation, the emotional counterpoint to injustice. It calls for artful outreach, with words, images, and facts chosen with a view to their simplicity, appeal, and potential to evoke sympathetic emotion in remote, unknown, and largely abstract audiences.

One significant outcome of these criteria for reaching or cultivating sympathetic audiences is an idealization of both victims and perpetrators, by which I mean their simplification to those essential qualities that reach a limited range of human emotions. Sarah Federman (2018) uses the common constructions of "perpetration," for example, to understand the contestations surrounding the French national railway's responsibility for transporting Jews to death camps in Germany.[10] According to Federman, the Société national des chemins de fer (SNCF) is in a certain sense an ideal perpetrator, with its inseparable connections to the symbol of trains: powerful, abstract in a way that makes it inhuman, and symbolically representative of the crimes attributed to it. The railcar thus came to stand for the horrors of the Holocaust. The ideal perpetrator is also strategically idealized, in that it has all of these qualities brought to public attention by a "champion opponent" or activist. "Labeling someone a 'perpetrator,'" Federman writes, "summons the 'essence' of perpetration they represent."[11] This last point is key: in many collective justice claims, as in popular fiction (again, pointing to the storytelling qualities of justice activism), perpetrators are known *only* by their essence, as abstract, powerful entities without human emotions, motivations, or frailties.

Victims, too, are idealized in justice campaigns, with qualities of innocence and vulnerability brought to the fore. Unlike perpetrators, victims are, to the extent possible, personalized, humanized, given faces, names, and stories. Sometimes the poignant suffering and death of a particular victim brings what can best be described as a global wave sympathetic horror in response to a humanitarian crisis. The 1972 Pulitzer Prize–winning photo of nine-year-old Kim Phúc, better known as the "napalm girl," by Associated Press photographer Nick Ut, as she ran naked down a road after a South Vietnamese napalm attack on her village, is widely credited with turning public opinion against the Vietnam War. The same kind of global sympathy took place in September 2015, this time attached to Alan Kurdî, the three-year-old boy photographed lying dead on a beach after drowning in the Mediterranean as his family tried to escape the war in Syria. Public sympathies that coalesce into global opinion tend to be directed toward children, and to a lesser extent the elderly, the innocent sufferers of violence—the "ideal victims."

Public outrage typically gathers naturally around violations that people can see, emotions that they can feel through visual cues, the sympathy-inducing heartbreak and pain of the displaced and dispossessed. The tears flowing down the face of a blue-eyed Rohingya boy in a refugee camp, captured by a photojournalist and sent out into the world, provoke the nurturing and sympathetic instincts that are basic to human nature. The tears of a black inmate (who was once a boy), arrested for drug possession through a predictive policing algorithm that creates an inescapable cycle of neighborhood-specific criminality, might not. He might just as readily be understood as a "perpetrator," his story of little sympathetic interest. The human rights violations that we are attentive to, that receive attention and compliance-inducing pressure, are simplified in ways that are *visual* and *emotional*, outcomes of media technologies and forms of consumption.

The result is a considerable limit to the number and kinds of violations that attract the kind of attention that, in turn, produces action and remedy in global legal institutions. Article 5 (1) of the Rome Statute, which lays out in broad strokes the mandate of the International Criminal Court, already restricts the crimes that are prosecutable in international law to "the most serious crimes of concern to the international community as a whole"—that is to say, genocide, crimes against humanity, war crimes, and crimes of aggression. The substance of these crimes, which are laid out in several articles that follow Article 5, read like (as in fact they are) a compendium of the very worst human behavior motivated by the most despicable inclinations, producing the most calamitous effects on the victims. So international criminal law already restricts prosecution to the most heinous offences. This restriction in law is consistent with the popular will that is the source of the law's legitimacy. It also raises the stakes of activism. For a claim to be heard, either by publics or by the custodians of international law, the crime of the state at its base has to be shown as egregious, horrific, and despicable to the highest degree. Under circumstances in which compliance ultimately follows from the pressures of public sympathy and indignation, the priorities and powers of international tribunals have shaped into something like a criminal justice system that prosecutes only serial killers. Those who commit the most odious crimes, which receive the most attention and provoke the greatest revulsion, are given prosecutorial priority, while just about everything else escapes the net.

Opponents of human rights have also used poignant examples of illegitimate suffering and death to *counter* the claims of humanitarian activists, using *fear* as a source of emotional contagion. White supremacists and nationalists have zeroed in on specific cases of the rape and murder of innocents by migrants as a way to heighten and exploit the anxieties associated with the influx of refugees fleeing wars and economic crises in Syria, Iraq, and Africa. This has involved reverting to the primordial instincts of stigma, hatred, and exclusion associated with perpetration. Matteo Salvini, in his capacity as interior minister of Italy, used this technique particularly effectively, amplifying specific crimes to compliant media outlets to justify sweeping policies that target all migrants.

In political arguments that use fear as a starting point, the refugee represents the ideal perpetrator, standing in for the many who collectively represent difference and danger. The "victim" in one context is recast as a "perpetrator" in another, as someone who represents visible difference, someone who is alien, threatening, and polluting. The tyranny of the single instance, the case selected to amplify anxiety as a way to shape opinion, reduces the complications and ambiguities of mass conflict to a single, digestible, relatable—and politically manipulatable—tragedy. Social justice activists promoting human and environmental rights are pitted against *antijustice activists* whose ideals of national integrity and white supremacy would have little chance of succeeding without their rejection of democracy, science, and facts as a foundation of the rule of law.

This means that in facing the task of persuading publics of the need to act against mass violence, human rights lobbyists have their work cut out for them. Diametrically opposed tendencies make use of different images of victim and perpetrator. Human fellow feeling based on conceptions of universal justice is in a state of open warfare with racist and nationalist tribalism that seeks to restrict feelings of sympathy and common humanity to those who belong, who are similar, and who are loyal. The natural inclination to protect the innocent can be enclosed within in-groups by rhetorically distorting conceptions of victimhood and redirecting them in the service of political interests.

Let us now consider how the dynamics of justice claims and public sympathy are influenced by the new media ecosystem.

Algorithmic Exclusions

Any feeling of virtual sovereignty or perception that the internet is an egalitarian space is quashed when the mechanics of search engine algorithms are unraveled. A fast-growing literature is setting off alarm bells about the uses of algorithms and big data in public policy. The abuses of algorithms include the process for determining qualification for poverty relief, such as the US social assistance program Temporary Relief to Needy Families (TANF). Virginia Eubanks draws attention to the ways that such "technologies of poverty management" fail those they are intended to serve by being shaped around "fear of economic insecurity and hatred of the poor."[12] Predictive policing is also widely invoked as a way that ghettoized racial minorities are targeted by self-reinforcing cycles of crime frequency calculations, concentrated policing, and high rates of incarceration, which, in turn feed into joblessness and recidivism.

The basic process by which these algorithmically mediated policy failures occur begins with stereotyped definitions of marginalized categories or communities of people. As I mentioned earlier, the most basic way of putting this is well known to programmers: "garbage in, garbage out." If the data used to create an algorithm is wrong or biased, the conclusions that emerge from the "black box" of analysis will reflect or amplify the error.

The media that lend visibility to certain kinds of rights campaigns channel them into particular forms. In performing this function of selection, search engines, contrary to their early promise, are not egalitarian. They act as gatekeepers to the messages that activists seek to communicate to online publics. Without high placement in algorithm-mediated search results, websites that advocate on behalf of human rights claimants will go unnoticed. Asal and Harwood make this point with reference to "minorities at risk" (MARs), those under threat of mass violence. Even under circumstances of imminent catastrophe, they point out, "political mobilization via the Web is conditional, dependent on a MAR's protest Website achieving high rankings within engines' search results."[13]

The achievement of a high rank by a MAR or any other human rights claimant is not easy. Google's PageRank algorithm, the corporation's key proprietary tool to rank websites and search engine results, can be a source of algorithmically mediated bias, with important implications for, among other things, the information on conditions of injustice that reaches (or fails to reach) public audiences. Sergei Brin and Lawrence Page, cofounders of Google, explain the way that they met the challenge of developing a search engine able to produce reliable results in a rapidly expanding web. Simply put, PageRank is a technology that ranks the significance of web pages by analyzing the backlinks (or "citation history") that connect one page to another. A page will be prioritized in a search if many other pages point to it or if a few pages that point to it have a high PageRank. By making use of the link structure of the web to count and rank the connections to a particular web page, the algorithm produces, as Brin and Page put it, "an objective measure of [a web page's] citation importance that corresponds well with people's subjective idea of importance."[14]

The fact that PageRank reflects existing subjective ideas of importance, how-ever, means that, by its very structure, the search engine can sometimes propagate prejudices and stereotypes. Safia Umoja Noble demonstrates this point in her book *Algorithms of Oppression*, offering a dismaying overview of the ways that discrimination is embedded in computer code and algorithmic technologies. Her foray into search engine "glitches" that reflect racial stereotyping began with a search of the term "black women," which yielded the top-ranked result HotBlackPussy.com.[15] At one point, a public outcry followed from the fact that the results of searching the word "Jew" directed users to sites advocating anti-Semitism and Holocaust denial. Google posted an explanation of this phenom-enon, with the recommendation that users select search words like "Judaism" or "Jewish people," which have less likelihood of connecting to hate groups.[16] The problem applies equally to searches for factual information. Shortly after the 2016 US presidential election, for example, Google searches for "who won the popular vote" produced as the top-ranked result a conspiracy blog claim-ing that Trump had won it.[17] The list of such offenses could (and does) go on. Many of these "glitches" are now preemptively fixed or are quickly corrected by programmers in response to public outcry, but the problem goes further than the offense caused by certain insensitivities in search engine results.

The main takeaway from this problem is that the PageRank algorithm and other search engine technologies are specifically designed to process and reflect back *existing information*: the history of the links to a page, which situates it in a hierarchy of significance. It is only as good as the data going in. This history does not evaluate the accuracy, propriety, or fairness of the links to a page, and it can include the most egregious errors and oversimplifications. It forces us into the meanings and categories of the already influential, the indulgers of instinct, and the noisiest fringes. To raise a sane and dissident voice over this cacophony is certainly a challenge; to draw attention to a previously little-known injustice, perhaps closer to impossible.

The central significance of search algorithms for human rights campaigning is that they bring ever-greater attention to causes that have already captured public attention and exclude causes struggling for attention from the "fringes." The result is a widening gap between claims that succeed and those that do not. Breaking into the crucial first page of a search to garner public attention and political leverage is a rarely achieved goal, one that only the most organized campaigns around the most egregious, visible, indignation-inducing violations of rights can hope to achieve.

These limits to the ability of activists to draw attention to a cause contribute to the circumstances in international law in which those who commit crimes on behalf of the state all too often escape with impunity. Autocrats have sometimes avoided prosecution in the International Criminal Court under the cover of a compliant press at home, one that cultivates national outrage at the court's perpetration of insults to patriotic belonging and slavish repetition of lies fomented by the ethnic and/or political opposition.

But we don't have to focus on such extreme examples of media manipulation to find an emerging trend in the connection between authoritarianism and a compliant "free" press. The new relationship between demagoguery and state-controlled media has sometimes been termed the "frisbee press"—that is to say, a relationship in which heads of state or high-ranking officials strategically throw a story for the compliant media to catch and retrieve (or broadcast or publish). In its intimate connection with Donald Trump, Fox News may have transcended the genre by throwing its own frisbees, offering unsupported stories that then are consumed by the president, tweeted out, acted on, and fed back to the media in a cycle that produces the bubble inhabited by Trump and his base.

In an important article in the *New York Review of Books*, Christopher Browning, a historian specializing in the Holocaust, Nazi Germany, and Europe in the era of the world wars, observes a central difference between the fascism of the twentieth century and the new authoritarianism of the twenty-first. The new authoritarians have discovered that they can maintain a firm grip on power while keeping basic democratic institutions in place but under control. The erosion of an independent judiciary as a check on executive power is one step toward centralized power. Another is the cultivation of a media environment that elevates the political agenda of the ruling class or party above all others. This is not quite the same thing as censorship, although media repression is seemingly a growth industry in undisguised dictatorships. The new form of authoritarianism has no need for such crude tools of repression. "Total control of the press and other media is . . . unnecessary," Browning observes, "since a flood of managed and fake news so pollutes the flow of information that facts and truth become irrelevant as shapers of public opinion."[18] This, he argues, is different from the overtly repressive tactics of the fascist regimes of the twentieth century, but it is no less troubling.

The manufacture of doubt is the reverse side of the Janus face of new media. The "frisbee press" phenomenon is not limited to manufactured stories; it can also apply more generally to the issues that garner saturation coverage. Media attention to the most recent mass atrocities is overlaid with stories that might be tremendously important but that turn all the spotlights in one direction, leaving issues of immediate concern in the dark. This is not a temporary condition. The proverbial "free pass" because of temporary public fixation on urgent news-of-the-moment has become a permanent condition of the media environment.

The Numbing Effect of Numbers

The new technologies of global governance work mainly through indicators. In the area of human rights this manifests itself in quantitative measures of things like death tolls and numbers of women and girls raped, refugees driven across a border, children dead from hunger, buildings destroyed, villages burned, and bodies in mass graves. Of course, in the fog of illegitimate war, much of this is invisible, kept as secret as possible within the limits of the inherent visibility of mass violence. Faced with the challenge of intervention, the UN needs the scope of the problem to be understood with indicators, even rough ones, so that the appropriate resources can be allocated. Without affixing numbers to problems, they are invisible and illegible to policy makers.

It is the power of numbers to convey an aura of objective truth and lend credibility to assertions of fact that leads Sally Engle Merry to speak of the "seduction of indicators." Assessments based on quantification "appeal to the desire for simple, accessible knowledge and to the basic human tendency to see the world in terms of hierarchies of reputation and status."[19] This goes some way toward explaining the growing reliance on indicators in international law. It is inspired, Merry points out, by the expectation that numbers make violations more visible, thereby applying pressure on states and corporations to comply.[20]

This account of indicators in international law makes sense in the context of the formal administration of law. Indicators are particularly meaningful and reassuring to officials responsible for measuring compliance. They appeal above all to the people in the "community of nations" or the "corporate community" who want reliable information about state and corporate compliance with international law. But what happens when we go outside the realm of officialdom, to the citizens and consumers who are another source of pressure toward human rights compliance? How do nonexperts respond to the seductions of quantification?

One of the problems with numbers is that by themselves they do little to convince public consumers of information of, say, the egregiousness of an environmental abuse or the urgency of a humanitarian crisis. Outside of corporate or global governance officialdom, numbers have to be used with great care, if at all.

To illustrate this point, the world of advertising can again serve as a guide. In advertising, the creative process is separated from the media vehicles that are going to be used to deliver the story to the target audience. The talents and goals involved in the two processes are entirely different. The creative team speaks the language of story, emotion, or "neuro-associations" such as joy ("open happiness!"), nurturing (Puppies! Kittens! Babies!), play (pan to an SUV with surfboards on a roof rack, with young people dancing their way to the beach), sexuality depicted in ways that appeal to the visual orientation of the male libido (the dancing to the beach example will suffice here too), and, very occasionally, fear ("Do you have a loved one living alone who can fall any time?"). The creative team determines how they want their audience to feel, with colors, images, and words serving as the raw material of persuasion.

In a different section of the agency (and sometimes even in a different agency), there is a media planning team that speaks the language of numbers. Once the creative part is done, media planning is applied using algorithms and AI to find exactly where the viewers are with babies and aging parents or other loved ones and who are likely to be most susceptible to the message. The planning stage identifies this audience and calibrates its presentation. This team has to convince the client, with the use of numbers, that its money is being well allocated. The story that they're telling is quantified. This team has to convince the client that they're going to spend the money well, with the right ad campaign, the right TV spot, billboard, bus kiosk, social media platform or website. Money follows the numbers. The data has to make the argument that for *this* campaign, the client is going to make *this much* money.[21] This corporate advertising scenario is not so very far away from the implications of numbers as human rights indicators in the context of public consumers of injustice messaging. Numbers in this context have an instrumental, not an emotional, purpose. Their task is to convince decision-makers, not the public. As raw material for popular persuasion, their utility is strictly limited.

All of this is to say that overreliance on numbers can be fatal to the public outreach goals of justice causes. As Stalin is reported to have said, "To kill one person is a tragedy, but the death of one million is a statistic." (The grisly implication of this statement in the context in which it was supposedly uttered is that a dictator can kill people in the millions with impunity—the number will attract little notice, as indeed was the case at the time that Stalin engaged in his purges at the cost of as many lives as were lost in all of World War II, including some 14 million who passed through his forced labor camps, or gulags.)

Like all aphorisms of this kind, however, it only goes so far. It is still true that there is some persuasive effect that only numbers can provide. They give a general sense of the existence and scope of a crisis and can arouse indignation from this simple act of consciousness-raising. This is especially true when mass killing is given concrete temporal and human points of comparison, as in a recent headline in *The Guardian*: "Activists, wildlife rangers and indigenous leaders are dying violently at the rate of about four a week," which was followed by a bullet point: "See the names of all defenders who have died so far this year *here*."[22]

This one news item uses two ways to get past the numbing effect of numbers. The first is to connect the indicators with a shared experience of the flow of time. It breaks up the number from an abstract mass to a more individual event, something that can be grasped from a perspective that gets past the overwhelming emotional abstractness of a statistic and transforms it into digestible units that lend themselves to imagination, connection of the publics with the loss, and the possibility of storytelling. This technique also connects the number with time in a way that builds urgency, as in the trope "A child is dying of hunger every fifteen seconds. A donation from you will help *right away*." Those inclined to sympathy construct their own mental image of a suffering child (if the image is not already provided for them) and relate that image to the passage of time: another and another, just like the one they thought about in its mother's arms or alone on a makeshift bed, every fifteen seconds. Donate now.

A second way that news items get past the numbing effect of numbers is more directly personifying: to connect a number to individuals, to give the names of victims is another way to give a number personhood. A list of names connected to each person's unique story on the web makes it more directly possible to imagine what the number means. Thus, the story in *The Guardian* on murdered environmental defenders includes a long list of individual victims with links to articles that describe their accomplishments in protecting the environment and the circumstances of their deaths. One simply has to multiply the feeling of sympathy one has with one case by the numbers it represents. The totality of it all may be overwhelming, but personification makes it possible to approach a crisis sympathetically and to intervene with an awareness of the human cost of an atrocity.

Numbers, however, can interfere in this process of personification. A study by psychologists Deborah Small, George Loewenstein, and Paul Slovic found that donations to aid a starving seven-year-old African child declined sharply when her image was accompanied by a statistical summary of the millions of other children in Africa facing similar circumstances. The numbers themselves actually got in the way of the psychological process necessary to get people to act.[23] Simply put, without exercises in personification, publics lack imagination. They begin with sympathy and an urge to help but tend to see numbers by themselves as unemotional, abstract, and lifeless, and they have difficulty imagining the suffering that the numbers represent. They have to be helped along by exercises in personification, connecting otherwise distant events to an imagined, more immediate narrative.

Storytelling

The task of personification through a narrative that provokes the imagination is often done with storytelling. Publics need to be told about injustice through stories and images that reach their emotions directly. Numbers alone cannot do this. Even the sloganeering efforts of repeating a catch phrase, saturating public space with sticker graffiti, or pounding a point on the table in meeting after meeting go only so far without being connected to stories. There is still a residue of oral society and its reliance on archetypes and the art of memory in the storytelling of justice claims. Stories need to have something captivating, a hook or a handle that easily gets picked up, that serves as a reminder of a wider story and gives guidance for its telling.

Nonfiction documentary film is the genre that best captures some of the storytelling qualities of the publics' legal imaginations. Conditions of injustice and social movements of justice claiming have long been staples of documentary filmmaking. Regina Austin offers a serious account of this phenomenon, intended to raise awareness of the documentarians' craft in the legal community.[24] (In fact, she invites her readers in the law profession to act as "authenticating audiences.") Documentaries, she points out, are ideally suited to covering complex issues by shifting between local and national (and, we might add, occasionally international) scales. They are also part of the phenomenon of "visual legal advocacy," most commonly used in the form of video settlement documentaries, in which victims themselves visually articulate their claims with the goal of legal persuasion. It is interesting to note that in 2006 when her article was written, the ways that Austin was able to see technology as altering the critical landscape of documentary films were limited to the directors' commentaries and outtakes from the filmmaking process used in DVDs and on the web, which offered better critical understanding of the observational rhetorical style. As I will discuss below, this aspect of direct observation in documentary filmmaking has since become much more important in human rights activism. The idea of visual legal advocacy has been picked up by documentarians as a form of live narrative that transcends the edited storytelling properties of the documentaries. Live streaming of stories as they unfold is prolifically practiced in direct action protests and other records of disputes in "real time."

In all of this focus on visual advocacy, it is easy to forget the centrality of *storytelling* to public receptiveness of visual claims making. Again, in this area the findings of researchers in marketing have the jump on legal scholars. Stories and storytelling are now a central part of the way that marketers understand consumer behavior. In particular, blogs were a source of the insight that individuals were providing commentary on their lives and the lives of others in ways that formed around storytelling arcs. Leveraging those stories, finding out the archetypal foundations of how humans think and communicate, was an innovative new way to reach and motivate consumers. Marketing plotlines correspondingly invoke the images of "jester, creator, ruler, rebel, sage, hero, outlaw, magician, or some other archetypal primal form."[25]

The public appeal of a justice claim or cause can similarly be seen to follow the narrative arcs and archetypes of storytelling. An example can be seen in the saturation coverage given to the Khashoggi killing. Before this event, mainstream media outlets emphasized the supposedly "liberal" reforms in Saudi Arabia under Crown Prince Mohammad bin Salman, or MBS as he was better known in diplomatic circles. The Western media and its publics were unmoved by information about the activists who had been arrested, including the women who had lobbied successfully for the right to drive and then had been incarcerated once the reform was put in place. The story about women being granted the right to drive seemed to gain far more traction than the story about the imprisonment of the very women who lobbied for that right. The activists were not much different from other dissidents and journalists moldering in jail in so many parts of the world. Their story did not stand out, was not different enough. Publics also paid relatively little attention to Saudi involvement in the war in Yemen and the catastrophic famines that tore through the countryside, as famines do, laying waste to thousands, mostly children and the elderly. The numbers were not enough to convey this suffering. Nicholas Kristof, an opinion writer for the *New York Times*, opened an editorial titled "Be Outraged by America's Role in Yemen's Misery" with a plea to his readership: "The news about Brett Kavanaugh and Rod Rosenstein is addictive, but spare just a moment for crimes against humanity that the United States is supporting in far-off Yemen."[26] The bombing of a Yemeni school bus using a laser guided GBU12

Paveway II bomb manufactured by Lockheed Martin, which killed some seventy people, including forty children ages six to eleven who were out on a school trip, temporarily roused some journalists and their publics from their torpor, but not enough to have any effect on the wide, media-driven acceptance of the new Saudi government as a vehicle of reform.

Then came the killing of one man, Jamal Khashoggi, by a professional hit team sent in a private jet to the Saudi embassy in Turkey with the ostensible purpose of silencing this journalistic gadfly. In one cold-blooded murder, the criminality of the government was much easier to imagine than it was in the UN's statistics coming out of Yemen. Mohammad bin Salman's acronym MBS, in a meme that raced across the internet, now stood for "Mister Bone Saw." The regime's contorted efforts to manufacture a plausible story to explain the killing became part of the story. Corporations now had their reputations to protect, torn between Saudi money and the downward spiral of its reputation, and many ultimately made the decision to pull out of the Saudi government's gala conference, the Future Investment Initiative or "Davos in the Desert," an event intended to bring together business leaders, investors, celebrities and members of the mainstream media. Tech companies walked a fine line between the embarrassment of association with a criminal government that attracted global moral revulsion and the investment needs of their corporations, which rely on massive infusions of capital and the cultivation of global supply chain partnerships. Under these circumstances, as Cecilia Kang puts it, "it's nearly impossible to unscramble the egg."[27] It takes an offense of major proportions to interrupt Silicon Valley's reliance on repressive autarchies and provoke a shift in the direction of corporate accountability for human rights.

It also took an event that made state-sponsored violence understandable as a story, a classic *Game of Thrones*-worthy tale of a vindictive and inexperienced prince exercising new powers in the ill-advised assassination of an enemy. The killing of one individual in a way that fit into deep human structures of narrative did more to bring attention to wider patterns of human rights violations than any number of killings and arbitrary detentions.

This is an insight that has not been lost on institutions of global governance. Storytelling can be seen, for example, in the practice of using of celebrity diplomats as a way to promote the goals of international intervention. This practice goes back to the appointment of the Norwegian Frijthof Nansen to the position of high commissioner for refugees in the League of Nations in a desperate context in which huge masses of refugees appeared in Europe in the aftermath of World

War I.[28] Nansen brought his celebrity as an Arctic explorer—and by association the remarkable adventures that took place during his explorations—to the
diplomatic tasks associated with building the first agencies designed to assist
refugees in the postwar era.

Today, the link between diplomacy and the persuasive power of celebrities
is established practice. Actress Angelina Jolie, for example, headlines the star-
studded list of celebrities serving as UNHCR goodwill ambassadors, while a
much larger list of celebrities stands behind the work of UNICEF (drawn, perhaps, by the inherent appeal of children): Mia Farrow, Orlando Bloom, Lionel
Messi, among others—even the Berlin Philharmonic, which joined in 2007.
They are effective as advocates not just because they are individuals with particular talents but because of the accumulations of storytelling built into their
biographies. At a certain point, to the public actors *become* the roles they play.
The stories that actors tell through their art or that athletes tell through their
victories migrate through their public commitments and are made manifest in
the persuasions of justice.

Visual Legal Narrative

Now let us look at how the use of visual storytelling plays out in specific circumstances. Considering how visual evidence and storytelling are deployed in
particular campaigns of public outreach offers insight into the persuasive effects
of new ITs.

The activists of the Dakota Access Pipeline standoff of 2016–2017 created one
of those rare movements that broke through algorithmic, PageRank exclusion
and journalistic filters and created a sensation. They plugged into a global network of sympathy and solidarity that was in some ways analogous to the Zapatista
rebellion of 1994, with similar sources of injustice and reliance on transnational
networks. The Dakota Access Pipeline protest's main differences from the Zapatista rebellion were (1) a commitment to nonviolence; (2) public outreach,
premised on similar strategic uses of new ITs as a supplement to mainstream
journalism, facilitated by different, faster, farther-reaching technologies; and
(3) prolific use of livestreamed, strategically presented, visual proof of nonviolence, often in the face of police brutality.

The organization behind much of the online coverage of events, Digital
Smoke Signals, bills itself on its website itself as "Indigenizing Technology: Walking the footsteps of our Ancestors as we educate the world through e-learning,
social networking & Film-making."[29] The Standing Rock movement did not

rely on the mainstream journalists, who descended in droves and who, looking for a story in moments of boredom, often awkwardly and against instructions gave attention to the colorful ceremonies practiced in the camp by indigenous participants. The movement largely bypassed (or at least supplemented) journalistic preferences and filters by putting information directly onto the web via networks of solidarity.

Myron Dewey, a Newe-Numah Paiute-Shoshone from the Walker River Paiute Tribe and a founding member of Digital Smoke Signals, talked about his experience as a "community journalist and activist" to a small audience of students and professors at an event hosted by MIT's Comparative Media Studies/Writing program.[30] Dewey specializes in the use of drones for recording major events of social protest, with aerial footage to be used for the creation of what he calls a "visual legal narrative." He made use of more than thirty drones at the Standing Rock protest, because they were continually shot at or confiscated by security forces.

The Dakota Access Pipeline protests (also known by the hashtag #NODAPL) began in April 2016 when Standing Rock Sioux elder LaDonna Brave Bull Allard set up a camp near the Missouri River and a sacred burial ground as a site of spiritual resistance to the pipeline. Over the course of the following summer, the camp grew to thousands of people from tribes across the United States and a wide array of international supporters, including indigenous people from Canada and Latin America and Samis from northern Europe. From the outset, those gathered at the camp strictly adhered to principles of nonviolence. Activists even avoided referring to themselves as "protestors" out of concern that the word could easily be used to justify violent repression and instead used the term "water protectors," to refer specifically to the members of the Standing Rock Reservation assembled to protect the river and its nearby sacred sites.

National and international press coverage followed. So did security forces that were sent to contain the gathering and move construction of the pipeline forward. Arrayed against the water protectors and their supporters were forces from county, state, and federal law enforcement (including several FBI "infiltrators" who joined the camp) as well as mercenaries from the private security firm TigerSwan (described on its web site as offering "solutions to uncertainty" and as "a global consultancy for global risk management and mitigation").[31]

As Dewey explained it, there was a debunking effect to his use of visual evidence over live feeds, an effect in which it was possible to undo the narratives of those representing the security forces and "explain to people watching that we are not who *they* say we are." When police denied that drones were being shot down, for example, it was reasonably straightforward to produce drone footage showing an officer aiming his weapon toward the drone's camera, firing his weapon, followed by a sudden jerk that came with the impact of the bullet. When the police denied that they had used "flash/sound diversionary devices" (commonly known as flashbangs or percussion grenades) against the protestors, video footage could be immediately produced that demonstrated their use, with the unmistakable flash and bang of the devices exploding, supported by images of wounded activists and spent canisters that remained in the field after the event. And when one of Dewey's drone pilots, Nodapl Prolific, was arrested and charged with endangerment for flying his drone in public airspace (criminal charges for which he faced the possibility of a seven-year prison term), he was able to assemble a five-minute video of his drone flying and filming activities for use in his defense. The charges were dismissed.

Dewey later explained to me the key reason that the troops sent to Standing Rock had initially been comparatively restrained in their use of violence. "It was simple," he said. "We had a minimum of thirty thousand people on a live feed all the time. They knew that people were watching."

Under a recently elected Trump administration, the political climate turned against the resistance to the pipeline. The camp at Standing Rock was broken up by force, with Dewey's aerial footage showing military personnel tackling people in the camp and leading them away in handcuffs. The one concession the water protectors were granted was a reinforced construction of the pipeline as it passed under the Missouri River to reduce the risk of a waterborne spill.

Despite this limited success in terms of changing government and corporate policy, the use of media by the participants in Digital Smoke Signals illustrates how the combination of established forms of visual narration with the immediacy of digital/online transmission has been effectively put to use. The limits of specialized technical expertise were overcome through consumer products (drones, smartphones, etc.). A wider range of people gained access to the (increasingly transferable) knowledge needed to deploy the ITs of organized protest.

Once the photographers and filmmakers of the protest camp managed to reach a critical mass of online followers, they were able to use that audience to amplify their messages even more, creating conditions for new forms of visual legal narrative or visual legal advocacy.[32] The pressure of accountability increased with the number of viewers. The use of live feeds was particularly important as a tool of visibility and accountability in circumstances in which on-the-ground activists encountered resistance from police and private security forces. These circumstances of physical conflict provided opportunities for broadcasting arresting visual images that added salience to the cause. At the same time, the live feeds drew attention to police excess and illegality in dismantling the physical and human presence of dissent. In these circumstances, live visual legal narrative was important as a record of injustice, one that served as an immediate brake on violence, a convincing source of evidence in court, and a longer-term, more permanent contribution to the documentary filmmaking genre.

The visual account of the standoff was also part of a documentary film project, which resulted in *Awake: A Dream from Standing Rock*. The film was screened at the Tribeca Film Festival in 2017 and distributed by Netflix in 2018. An online minitrailer presents the following lines: "We are something new on the planet. But we are something very old. We are warriors of peace," a reference to the project's wider goals of outreach.[33] The film concludes with an argument about universalism, moving from the single instance of the camp at Standing Rock to other examples of protest against pipeline construction in the United States to a global movement of peaceful environmental advocacy on the model of the "water protectors."

Head and Heart

There is inherent appeal in hearing a story from its source, from those who convey the events and emotions of personal experience. Those immediately affected by human rights violations have long been understood as the most compelling advocates of policy reform and redress.[34] Observing this principle in practice, Winnifred Tate followed Colombian activists to Capitol Hill, where they met with policy makers to appeal for changes to US programs in their country. The practice of *testimonio* by the Colombian villagers, Tate observed, involved "particular forms of emotional expression as well as the presumed implication of the listener in political commitments through the newly acquired comprehension of suffering."[35] Members of Congress who were unaccustomed to hearing personal accounts of devastating tragedies, were overwhelmed with stories of villages burned, crops destroyed, and family members and friends disappeared and brutally killed. This, Tate found, constituted an unfamiliar form of discourse in Washington and a new strategy toward fostering solidarity between rights claimants and policy makers.[36]

At the same time, campaigns for justice do not succeed without meticulous documentation. Victim narratives told through tears are not enough. In fact, there could well be a turn away from the "relative truth" of such narratives in the context of their post-truth politicization, particularly in conjunction with the internet's capacity to convey attractive disinformation faster than more prosaic truths. This is particularly so of claims based in the experiences and narrations of historical trauma. Stories have to be supported with incontrovertible evidence before publics will invest their sympathy in them.

The obvious strategy, then, is to combine these approaches to persuasion, to appeal to both the head and the heart. The most convincing, incentivizing stories are those that make use of facts *and* narrative, that tell stories and reach emotions even as they engage in justice-oriented documentation. The ideal strategy involves the construction of a "file" that is persuasive both in court and in the court of public opinion. But it also involves storytelling. We are creatures of language and thrive on evocations of imagination. If you are a dissident, there is a lesson in this: facts are great, but if you *really* want to reach people, look beyond numbers and imagine a campfire.

CHAPTER 5

Media War

Pauvre, pauvre Imollen	Poor, poor Imollen[1]
Viens	Come
semons nos pas	Let's sow our steps
dans la poussière des sentiers	in the dust of the trails
Ces voix qui t'ont distrait	Those voices that troubled you
cette nuit	tonight
sont seulement complaints de	are only complaints of
l'agonie	agony
gémissements du deuil	groans of mourning
chants et larmes du désert et de	songs and tears of the desert and
la soif	of thirst
menues offrands au silence	small offerings to silence
pour engendrer d'autres jeunes	to generate other young
vents	winds
qui s'enflammeront demain	that will ignite tomorrow

Hawad, *Furigraphie: Poésies 1985–2015*[2]

The WhatsApp Revolution

While I was talking with a Tuareg activist in a cafe in London, his iPhone constantly lit up and buzzed with incoming messages. Mahmud and I had met a few times previously in New York at UN meetings that he attended with the sponsorship of the Indigenous Peoples of Africa Coordinating Committee. Now we were in the Chelsea district having lunch in a hip Asian-fusion cafe and talking about terrorists, human rights, and his work as a tech-savvy promoter of a negotiated peace for the central Sahara. He seemed almost proud to tell me that he was "number one on the hit list" of al-Qaeda in the Islamic Maghreb, and for a fleeting instant I couldn't repress the thought that, while we were together at the same table, so was I. By way of excusing the distraction from his blinking, vibrating iPhone, he showed me his WhatsApp broadcast list, consisting of those committed to an independent territory in the central Sahara, which seemed to never end as he scrolled through the profile images of mostly turbaned men. His network included family and friends from across Europe, North Africa, and the landlocked countries of West Africa (Mali, Niger, and Burkina Faso), including people herding goats and camels in the remotest parts of the central Sahara.

WhatsApp has had a powerful effect in bringing together Tuareg migrants in Europe with common language and shared experience. In regions where mobile service has reached the desert, Tuareg and Berber pastoralists, who travel as far as 400 kilometers with their animals across forbidding landscapes, are calling home and forming far-flung networks like never before. Even in circumstances of spotty connectivity and often lacking literacy skills, they are able to communicate using the calling and voice-recording features of WhatsApp. This goes some way toward explaining the platform's popularity. Equipped with smartphones, they are now able to hear familiar voices from the backs of their camels or as they shelter from the sun, break for tea, and set camp. When members of the Tuareg diaspora, some of them recently arrived refugees who had crossed the Mediterranean in rubber rafts, wanted to find their way to an out-of-the-way village in France for the annual meeting of the Organisation de la Diaspora Touarègue en Europe (ODTE, or Organization of the Tuareg Diaspora in Europe), they used WhatsApp to coordinate their movements, find lodging, share transport, and ultimately make the two-day event possible, despite their limited means.[3] The annual meeting brings together people living in Italy, France, Spain, Belgium, and Germany, some more permanently settled than others but all of them tuned in to their rights as migrants, refugees, and members of an indigenous people of the central Sahara in search of a homeland.

Among its many impacts, smartphone technology is creating new networks that cut across what were once caste-like tribal boundaries. To put this in context, we should bear in mind that feuds between rival lineages resulting in bloodshed were taking place in the central Sahara as late as the mid-1980s, when I was living in Gao, Mali. One of the Qur'an schoolmasters with whom I had regular contact explained his absence for a period of several weeks by telling me he had been summoned to negotiate peace between members of two tribal groups who were fighting one another with swords. Consistent with this observation, Jeremy Keenan reports an experience he had while making his way to a remote region of southern Algeria in which his travelling companions from rival Tuareg lineages were unable to conceal their hostility toward one another, and as the trip wore on, "insults and accusations began to dig deep into the Kel Ahaggar-Kel Ajjer war of 1875–78," the dates of which they had forgotten but the major events of which they remembered as though they had taken place yesterday.[4]

The use of WhatsApp in the context of forced diaspora has significantly broken down attachments to these kinds of tribal divisions, or at least structures new relationships that are quickly forming under real-world pressures. The tribally overlapping contact lists on social media and the common concerns being expressed through them give more permanent reality to an umbrella identity, that of the "Kel Tamasheq" (the Tamasheq-speaking people). This identity, in turn, has added greater significance and coordination to claims of recognition, peace building, and regional autonomy in a territory of the central Sahara that the Tuaregs call Azawad—their homeland and the focus of their claims of self-determination and human rights.

Lest I present too rosy a picture of these developments, Tuareg groups oriented toward violent secession, most notably the Mouvement National pour la Libération d'Azawad (MNLA), as well as more extreme groups affiliated with al-Qaida and ISIS, including the drug traffickers that finance these organizations, have been making use of this the same technology in the central Sahara. They are driven by entirely different values but overlap with the same social world of the diasporic, liberally minded migrants and refugees. WhatsApp's encryption features in particular allow them to coordinate their movements untraced in a sporadically violent standoff against UN peacekeepers and French and US troops who are leading military efforts to control the region, with only limited and sporadic success. Owing to this low-level guerilla-style insurgency, the approximately 11,000 peacekeepers living behind security checkpoints and fortifications in the Gao region[5] have been assigned to the deadliest UN mission in the world.[6] Social media platforms facilitate and amplify values that are diametrically opposed, while being cultivated among a people who share essentially the same history and language. There is also a practical side to encrypted social media that enables the actions associated with these values: on the one hand, it facilitates the creation of bonds of pan-European migrant solidarity and lobbying in the UN's NGO meetings in pursuit of recognition and regional sovereignty, and on the other hand, it is used by those responsible for blowing up UN peacekeepers and civilians who happened to be in the wrong place at the wrong time.

In this chapter (and the next) I situate new information technologies in a complex case study that draws our attention toward those who are struggling for rights and recognition through legal pathways that rely on campaigns of public persuasion. To bring out the difficulties of human rights claims making by those who are the subjects of gross violations of rights and dignity, I situate their acts and artifacts of persuasion in the context of shifting political alignments and histories of violence. As soon as we leave the comfortable predictive technologies of law and of information and their immediate effects, we are thrown into a wilderness of political struggles that shift and disorient like dunes in the desert wind.

There is a basic lesson to be drawn from this exercise: new technologies have not solved a basic problem in the foundations of human rights claims making, which follows from the fact that publics are limited in their ability to grasp and tolerate complexity. Publics tend toward stereotyping, misapprehension, and oversimplification. They have difficulty imagining a distinct people as anything other than a uniform, bounded entity, with a coherent structure of decision-making and shared political aspirations. What is more, a collective rights claim gains no traction in public opinion, and hence has no effect in changing conditions of oppression, if the public on which it depends acquires a perception of a flaw, or a failure to correspond with an ideal, somewhere (anywhere) in a factionalized society. It can—and did in the case of the Tuareg—even fail to respond to a serious crisis by casting the human rights activists and peace brokers in the same mold as violent insurrectionists. As I will show, new communication platforms have not accommodated a public understanding of (or even tolerance for) this complexity but have produced and reproduced stereotypes that lay at the base of both the sympathies directed toward victim-claimants and the narrow structures of knowledge that keep them at a remove from both their homelands and redress of the injustices they experience.

The Tuaregs arguably represent widest range of value differences of any one people, with individuals faced with choices that include commitments to violent jihadism,[7] waging war as a means toward self-determination, and human rights advocacy. For those who are committed to the legal pathways of justice, regional autonomy, and national reconciliation, this situation presents a logistical nightmare. With those who aligned themselves with Al-Qaida commanding international attention, the claims of justice lobbyists are invisible and unintelligible to publics, and their conceptions of human rights and regional autonomy are ignored. Even though supporting those with commitments to international law is a promising strategy for combatting terrorism, stereotypes and futile military options prevail.

Through the events of the 2012–2013 civil war in northern Mali and its aftermath, a picture has emerged of a starkly divided society, with media depictions of violent insurrection taking up the foreground. One strategy pursued by the Tuaregs toward recognition and autonomy calls for effective use of the tools of persuasion, using computer literacy, media savvy, and legal knowledge (particularly knowledge of the inner workings of UN agencies) to mount challenges to the political and developmental ambitions of states, often by calling into question their record of rights compliance as members of the international community. They pursue this strategy in common with a wide range of human rights activists from many parts of the world, making it a central aspect of their participation in the global indigenous movement. Among the Tuaregs, this strategy is pursued by a relatively young, educated elite, a computer literati that is able to effectively navigate both the bureaucracies that sponsor rights claims and the media technologies necessary to convey these claims to members of their communities and to distant, potentially sympathetic consumers of information. In the global indigenous movement, this calls for making a case for regional political autonomy and control of territory as necessary for the pursuit of a subsistence-based way of life. This in turn entails an argument to the effect that the collective knowledge (sometimes referred to in global governance circles as indigenous knowledge [IK], traditional knowledge [TK], or traditional indigenous knowledge [TIK]) of these collective rights claimants constitutes not only a birthright for those who belong but also a wider contribution to human heritage for those who do not.

The other strategy calls for an effective use of politically motivated violence, with leaders who are especially knowledgeable about desert life and its conditions of war, using the alliances necessary to create a force capable of effective action against the military presence of the state. With fewer fighters and lighter weapons than the states' militaries, this strategy involves use of mobility, surprise, and (ultimately) the psychological effects of spectacular violence and demoralization to wear down state militaries and UN peacekeeping forces and drive them out. In the longer term, violent resistance can then be converted into a kind of negotiating capital with which the conditions of an agreement of regional autonomy with the state can be (so it is sometimes hoped) advantageously formalized. The

Tuaregs' most recent war of independence in northern Mali in 2012–13 and the failure of the subsequent peace accord of 2015 reveal that the politics of power can occur in parallel to the politics of shame, sometimes with the same goals of self-determination and regional autonomy in mind.

Human rights lobbying often occurs in institutional contexts in which any recourse to violence is morally repugnant. The use of force and, above all, the cultivation of alliances between Tuareg Islamists with insurgents based in Libya, not to mention their association with an internationally recognized terrorist organization, Al-Qaida of the Islamic Maghreb (Al-Qaida au Maghreb Islamique, or AQMI), severely undercut the human rights claims coming from more moderate Tuareg organizations. It did this above all by undermining the commitment to nonviolence as a precondition for participation in human rights forums and for demonstrating the kind of cultural merit that is a foundation of public sympathy and support of rights claims. Although the Tuaregs' NGO representatives, by virtue of their commitment to nonviolence, were permitted (even encouraged) by the rules of the UN to continue their work, they had to struggle against media reports that connected their people with a brutal war in the central Sahara, producing an impression of two-facedness that was not of their making.

The sharp contrast between the moral orientations acted on by different Tuareg groups raises a simple question: how could one people present themselves openly in human rights forums as victims of state-sanctioned violence and dispossession, while at the same time (so it would appear to a media-consuming public) forging alliances with insurgents and terrorists in efforts aimed toward the violent occupation of the central Sahara? There are also more difficult issues raised by this case: how is the implementation of human rights possible where those rights have little, or inconsistent and divided, currency among the potential rights claimants? Further to this, how, in morally complex struggles for justice, are some states able to exploit the apathy and divided attentions of publics in ways that systematically, and at times egregiously, violate international justice standards? And what might be the place of new ITs in cultivating stereotypes and entrenching conditions of rights violation and violent conflict?

These questions turn largely upon the uses of media as pathways to justice. The difficulties faced by Tuareg rights lobbyists in sympathetically presenting their cause in the context of a wider violent insurgency can serve as a starting point for considering a number of the more general qualities and consequences of media-centered collective justice lobbying. Before turning to this issue, however, I describe in some detail the causes and conditions of the Tuareg insurgency.

To bring home my wider point about the structures of media creating grossly oversimplified pictures of the reality facing justice claimants, I first have to depict the conditions of violence in Tuareg society in a way that does justice to their complexity. This will make it easier to see the entanglements and dilemmas of public outreach by those who do not comfortably fit within the accepted repertoire of human rights victims.

Three Models of Resistance

Even though the coherence of Tuareg lineages has weakened and even though the differences between social strata have seemingly kaleidoscopically shifted ever since the earliest French occupation more than a century ago, there is still a political dimension to both tribal affiliation and social status that acts upon decisions of war and peace as well as participation in NGOs and international forums.[8] This has resulted in a shifting scene of political divisions in which three basic models of resistance to state power have taken form, each with their own approaches to public persuasion through new media.

In this resistance complex, no individual person with charismatic authority and no organizational or tribal entity represents the Tuareg as a whole. No single leader, no lineage, and no community of paramount chiefs controls the various military alliances, Islamic orientations, or internationally sanctioned nongovernmental organizations. Each of these political entities has very different collective aspirations and sources of legitimacy, yet all are juxtaposed uncomfortably within the ethnonym "Tuareg." The terms I use for these models, using the suffix *-ism* to indicate their deep politicization, are *secessionism, jihadism,* and *indigenous NGO-ism.* Let me discuss each in turn, with particular attention to their origins and online presence.

Secessionism

For those who might be tempted to assume that secessionism in the central Sahara might be easily resolved through a negotiated solution, there is not only the involvement of five states to consider, but perhaps more significant, a deep history of violence between the Tuaregs and governments with ambitions to control their world. Despite the elemental harshness of the desert environment in which they live, the most dangerous forces that challenge the survival of the Tuaregs come from outside the desert, largely beyond their reach and reckoning. For centuries, they have been subject to, resistant against, and sometimes fractured by the power of states and civilizations that surround them. They live in the central Saharan and northern Sahelian region of West Africa, a region that was long the gateway between Maghrebian, Arab, and West African influences. In the early colonial period, they were impacted by the Ottoman, French, British, and Italian troops that divided amongst themselves the various mountain regions of the central Sahara. The economically motivated expansion of European interests occurred through the imposition of new colonial frontiers, severing the once flexible, permeable, shifting boundaries between the various groups of desert pastoralists.[9]

Some aspects of French colonial rule were responsible for imposing profound changes to Tuareg society, which contributed to the nomads' inclination to revolt using guerilla tactics in the inhospitable, mountainous regions of the desert. Other aspects, particularly economic and educational policies, encouraged sedentarization and integration of nomads into the institutions and values of the state—and access to the communication skills and technologies acquired through state education.[10]

With the independence movements of the 1960s and the establishment of independent African states through most of the continent, the political fracturing of the desert region by foreign powers became even more acute, taking the form of patrolled frontiers, defended by symbols and political stratagems of the newly minted modern states that aspired toward control of their remote territories. Under these circumstances, the Tuaregs found themselves divided between five distinct states—Niger, Algeria, Libya, Mali, and Burkina Faso— with frontiers drawn on a map with little regard for those people who subsisted as camel-herding pastoralists and who migrated over vast distances across the invisible state borders of the desert region. Corresponding with these five states were different school systems, competitive economies, and mutually exclusive, sometimes hostile, political ideologies that took over the colonial borders while promoting European models of states and democracy and maintaining their centers of power in capitals far from the central Sahara.[11] As a result, the nomads of the desert today occupy a periphery on the outskirts of a periphery, with the administrative centers of the northern region—"outposts" like Timbuktu, Gao, and Kidal in Mali and Agadez in Niger—being widely known as frontier towns, remote from the ambitions and political attentions of the government leaders in the south.

The one thing the states that divide the Tuaregs have in common is a lack of legitimacy and the imposition of policies that have occasionally sparked violent insurrection. A Tuareg uprising in Mali in 1990, against a backdrop of drought and settlement policies, for example, was marked by a cycle of revolutionary violence, state-sanctioned repression, internecine warfare, negotiated peace, and bad faith, followed by simmering resentment, disaffection, and the return of a proclivity to revolutionary violence. The 1990s uprising ended first in the short-lived Accord de Tamanrasset of 1991, then in an Algerian-brokered peace treaty, the Pact Nationale, of 1992. On paper, this latter treaty provided the Tuaregs with some degree of self-determination in northeast Mali, and the government in Bamako committed itself to integrate rebel leaders into the Malian army and administration. But the agreement soon succumbed to a lack of national political will, while on the part of the Tuaregs it led to a situation aptly described by one observer as "a bitter soup of acrimony and acronyms as the [Tuareg movement] split along ethnic and tribal fault lines."[12]

Once the peace agreements failed, the state-sanctioned violence continued. More than six hundred extrajudicial executions of nomads took place in Timbuktu between April and June 1994, a prelude to further systematic killings in the neighboring district of Gao.[13] Not only were government forces credibly accused of such atrocities, but the government also supported militias organized around the specific goal of targeting Tuaregs, notably the Ganda Koy (masters of the earth) movement of the Songhay agriculturalists, with violence driven by ideals of ethnic purity.

These conditions of conflict were the political wellspring of the Mouvement National pour la Libération de l'Azawad (the National Movement for the Liberation of Azawad, or MNLA). The social media posts of MNLA followers are entirely consistent with what one might expect from a militaristic movement of secession: youths in white T-shirts with the logo of the organization do parade drills next to the flag representing their vision of the independent state of Azawad; Toyota Hiluxes mounted with 50-caliber machine guns drive past the camera along a desert road, with an occasional close-up of the young soldiers in the cargo area clinging to the gun mount; and the main square in Kidal as a crowd moves into a tight knot, fists raised, and chants "A-za-wad!" at the arrival of their leader, Bilal Ag Acherif—all the ingredients of the kind of national loyalty needed for a protracted war.

Jihadism

In March 1996 a flame of peace (*flamme de la paix*) was lit in a ceremony in Timbuktu, in which some 3,600 rebel arms were destroyed, with the melted remnants of the weapons later used to construct a peace monument. But the resentments that simmered after the conflicts of the 1990s were too great a force to be contained by a mere symbolic representation of peace. History repeated itself in May 2006 with an uprising led by Iyad Ag Ghali against the military garrisons in Kidal and Menaka, soon followed by another regime of peace, this one referred to as the Algiers Accords. In furtherance of this negotiated peace, Ag Ghali was assigned a position as Mali's consular representative in Saudi Arabia in 2007, but he was expelled three years later because of suspected connections with Al-Qaida. In 2012 he became head of the Islamist organization Ansar ad-Dīn (defenders of the faith), one of the Tuareg groups that in the same year initiated a war for an independent state of Azawad. This violent attempt at secession prompted a hastily assembled French military alliance with the government of Mali in early 2013, which soon achieved the reoccupation of the north and the military defeat of both jihadism and Tuareg separatism, converting a rebel army advancing on the capital Bamako into an insurgency, with an uneasy truce punctuated by sporadic violence toward French and UN peacekeepers.

Salafi-inspired Islamic reform, which found its way to the central Sahara by an extensive variety of influences, must also be counted among the forces that have had the effect of crosscutting lineage loyalties and unsettling traditional forms of leadership in Tuareg society. Much of the attention given to the Salafist organization, the Ansar ad-Dīn, which coalesced around these influences, is focused on strategic and military considerations, above all the place of the organization in the shifting forces and alliances that marked the secessionist occupation of northern Mali from 2012 to 2013. But another, largely overlooked consideration involves the public representation of Islamic reform and the impact it had on shaping opinion, above all on making opinion in Europe and North America unfavorable toward or unreceptive—in the sense of not even being able to perceive—claims of rights and regional autonomy in the hoped-for homeland of Azawad. In this sense, the International Criminal Court's (ICC) conviction of Ahmad al-Faqi al-Mahdi in 2016 (discussed in chapter 3) could well have unintentionally contributed to the stereotype of Tuareg = jihadist. The high-profile platform of the court and its use of open source intelligence to analyze social media depictions of violence amplified the image of jihadist violence, contributing to the way that it came to represent the Tuaregs as a whole.

For their part, the media posts of the Ansar ad-Dīn, had their greatest impact during brutal armed struggle for the autonomy of northern Mali in 2012–13, which jolted viewers in southern Mali with images of the cruel execution of the military leadership captured in an abortive counterinvasion. The insurrections that have periodically flared up during the past several decades tend to involve guerilla tactics, hand-to-hand combat, and mass killings, the kind of close-quarters violence that evokes fear among soldiers in the remote outposts who are targeted by the assaults and disgust among those who hear about them, and that also provokes revulsion in circles of international diplomacy.[14] They have since shifted toward the global model of jihadist violence: roadside IEDs, car bombs, and suicide attacks. These are among the themes that appear in the social media posts of Ansar ad-Dīn. Jihadist pride is manifest above all in military prowess: fighters lined up, lying prone along a sand-and-stone embankment, machine guns at the ready; smiling portraits of comrades in arms wearing army fatigues and AKs; and a result of this celebration of violence that I found posted online: a photo taken from above the body of a high-ranking Malian army officer, his throat cut, blood emptied in the sand. Then there are the images of their leader, the man who proudly takes responsibility for these acts, Iyad Ag Ghali bearded, smiling, with an ample white turban, and in every image surrounded by his adoring people.

Indigenous NGO-ism

Against the dominant narrative of the postcolonial state, the Tuareg's NGO-based leadership faced the challenges of trying to appeal to public audiences and, through public support, create space for local and regional autonomy. This was impossible in Mali, with government-controlled media and a national slogan—one people, one goal, one faith (*un peuple, un but, une foi*)—serving as a reference point for media campaigns emphasizing national "solidarity." In other words, Tuareg representatives, in common with those of many other indigenous people, were compelled to navigate a new moral and media landscape of oppression. In their participation in UN meetings and social media posts, they emphasized their marginalization in the project of the state and represented their cultural virtue and the legitimacy of their human rights claims to a wide, mainly European audience attuned to issues of social justice and cultural preservation.

There is certainly raw material in the traditional knowledge of the Tuareg to support a sympathetic approach to their distinctiveness. Susan Rasmussen, for example, has written on the spirit possession practices of Kel Ewey women, with sickness understood to be the work of malevolent spirits and, in response, with the singing, clapping and drumming of possession songs linked directly to cures: "It is said that the spirits are pleased by the drum calabash, ululation, applause, and the tonal quality of the music."[15] Consistent with this kind of appeal, international touring groups such as Tinariwen and Bombino have received huge media attention through their ability to effectively combine music and rights claims, with stage performances that include accounts of the Tuaregs' obstacles to prosperity and chronic conditions of injustice.

In European public opinion, originating in particular from France, Switzerland, and Germany, there is also a long-standing positive attitude toward the Tuaregs that is somewhat forgiving, if not appreciative, of their inclinations toward violent insurrection.[16] After independence, as personal encounters with Europeans of various nationalities increased through tourism, development initiatives, and marriages, more opportunities arose for closer contacts, strategic alliances, and sympathetic perceptions.[17] In keeping with this appreciation from afar, a spate of books connected sympathetically with the long-standing struggle that Tuareg insurgents have engaged in with the French colonial empire and modern African states, books in which they readily associate their armed struggles with the romance of difference. This theme is captured, for example, by the title *Touaregs: La révolte des hommes bleues 1857–2013* (Tuaregs: The revolt of the blue men), the color blue being a reference to the indigo dyes commonly used in Tuareg clothing, a striking symbol of both cultural difference and adaptability since the dye serves as a natural barrier against sun exposure.[18] The appeal of their capacity for survival in the desert is thus associated with their proclivity to "revolt."

The romance of difference is inseparable from NGO outreach and claims of autonomy. The hoped-for homeland of Azawad, often coupled with the romance of the desert, is a regular theme of Tuareg activism in Europe and a reference point for representation online. Young people in particular tend to question the state borders that have cut across the Sahara, sharing among themselves a poignant sense of their illegitimacy. They are the ones who mount the flag of the independent homeland of Azawad behind the speakers at every meeting of the Tuareg diaspora in Europe, taking care to make sure it is straight, like a carefully hung picture: "A little to the right. No, too far. Good. Down on the left. Stop. Up a bit. Perfect!" At one point during the 2018 annual gathering of the Organisation de la Diaspora Touaregue en Europe, a group of young men literally wrapped themselves in the Azawad flag, taking selfies that they then posted on social media, happy moments instantly shared across borders and continents (figure 5). During these annual gatherings, smartphones are ubiquitous and every poignant moment is subject to saturation social media coverage. The imagery they emphasize in Facebook posts goes to the heart of both their nostalgia for a distant homeland and the romance of desert life that finds favor in European audiences. Turbans and clothes with finely woven fabric (some traditionally indigo-dyed), impromptu music with clapping following the rhythm of a calabash, dancing with movements that convey modesty and grace, a "fashion show" displaying not only the art and beauty of Tuareg couture but implicitly the slowness and patience needed to produce it—these events bring out smartphone cameras with the intensity of press scrums. Participants' attentions to the activities before them alternately tune in and become absent as images are captured and then posted online.

Behind the pride expressed in images taken and posted to audiences in Europe and Africa, there is also hope for durable peace in the central Sahara. In fact, though youth may express it more boldly, the achievement of a pan-Saharan homeland is a common ideal, one that reaches across the generations of Tuareg society. The states' borders that were drawn and increasingly enforced in the postindependence era are a foundation of political grievance and a central source of the will of the most disaffected among the pastoral peoples to achieve autonomy for the central Saharan region.

The Tuareg NGO social media posts and online activism have accordingly emphasized indigenous internationalism. In their outreach, the Tuaregs' NGO activists have campaigned on a platform of transnationalism, peace, and the

Figure 5 Selfies, draped with the flag of Azawad (right).

suffering of displacement. If the pursuit of the autonomous homeland of Azawad is there, it is usually in the background and framed in the context of indigenous rights. There are plenty of images of indigenous NGO representatives from different parts of the world smiling at the camera, embodying their commitment to diversity with embraces and faces close for the pose. There are announcements of significant events such as the annual meeting of the Permanent Forum on Indigenous Peoples or the General Assembly's launch of the International Year of Indigenous Languages. The posts put Tuareg actors in the thick of things, at the microphone in moments of political gravity, leaving viewers to read the captions to get a sense of the words caught by the camera in mid-formation.

There are also less illustrious events online, organized at the community level. In a meeting of the diasporic community in Paris, video-recorded for posting on social media, the poet Hawad held the floor, interrupting other speakers, gesticulating flamboyantly as a way of adding emphasis to his words. Switching between French and Tamasheq, he spoke of the suffering (*douleur*) of exile and pride in a people whose timeless tradition of mobility gave them claims to an identity that stretched as far as their movements, from Egypt in the east to France in the north. As in the epigraph to this chapter, his words were sorrowful, nostalgic, and hopeful, with the "songs and tears of the desert" still pointing to the "young winds" of tomorrow. Among his audience were a group of youths, listening in rapt attention, and rights lobbyists at the table, appearing slightly annoyed at their inability to get a turn at the microphone. One apologized to me on Hawad's behalf after the meeting. "That is how it goes with us," she said. "We have to be patient." Then after a pause she added, "Every independence movement needs a poet."

Leadership Differences

This spirit of nationalism has gone together with an acute sense of injustice that has served as a defining quality of collective being and acts as the motivating force behind the Tuaregs' justice claims and lobby efforts. From the beginning of the African presence in the international movement of indigenous peoples in the 1990s, the Tuaregs (in common with and supported by the Berbers, or "Amazighs," based more in north Africa) asserted indigenous status, pointing to their collective experience with forms of oppression held in common with other indigenous peoples from many parts of the world.

The Tuaregs' NGO activists, however, are in an uncomfortable position from which to claim regional autonomy. Not only are the popular ideas about the

Tuaregs in some ways contradictory—an almost inevitable consequence of taking particular qualities and practices out of context and making them part of a timeless inner essence—but there are at the same time fundamental differences between the leaders who express and in some cases embody these ideas about who the Tuaregs really are. The divergent narratives of their online presence that I have just outlined reflect structurally entrenched political differences. Those who control the politics—above all those who make decisions about strategic alliances and about war and peace—tend to be different from those who control claims of rights and collective representation of the Tuaregs as a people whose rights have been systematically violated. There is a dual origin to militant action that resides partly in the differences between generations, between (admittedly putting it too starkly) youth leaders who are formally educated and tribal lineage leaders whose knowledge is based in desert life. This intergenerational difference is very common in indigenous societies in many parts of the world, but among the Tuaregs it is especially noticeable.

The difference between literate activists and nonliterate militants results first of all from social differences relating to pastoralism and the power of desert mobility. Ultimately, it has to do with pasturage and the differences between long-range mobility and greater sedentarism and between the military advantages of mobility versus access to the institutions of the state through settlement. This in turn has had implications for educational opportunities, and hence for the leadership acquired from knowledge of law and the communication skills needed to put legal knowledge into the form of publicly mediated activism, including social media posts and justice campaigns. In short, there is in Tuareg society a complex interplay between lineage organization (and its partial dissolution), historical patterns of transhumance, formal education, and a divide between a comparatively young, educated, computer-savvy leadership and the traditional political authority that determines the choices of war and peace. The model of scattered groups practicing "ecological nomadism" might correspond with the ideals of early ethnographers and of more contemporary, ecologically attuned non-Tuareg rights activists, but it excludes the sedentary dimension from the analysis of Tuareg society.[19]

The altered pattern of social strata and mobility, in turn, had an influence on the values and opportunities related to French education. In some cases, elite families, faced with state demands to send children to boarding schools, preferred to send daughters rather than sons because the labor of sons was needed in the families' patterns of transhumance. For similar reasons, many former captives or their descendants filled the quotas for school attendance imposed by French and later state-initiated universal primary education programs oriented toward sedentarization and assimilation. The result is that many Tuaregs whose leadership is derived from formal education came originally from among those who settled closer to urban centers. Gael Baryin reports that in the 1970s, as few as twenty Tuaregs from the desert regions of northern Mali went to secondary schools in Timbuktu and from there to higher education in Bamako, and of these almost none were from the major tribes (*grandes tribus*) of the central Sahara.[20] The educational preference that gave an advantage to women and the *kel aghrem* was to have long-term consequences. The most significant was a disconnection between those whose leadership was based in the politics and power of life in the desert and those whose leadership followed from education and rights lobbying in civil society organizations.

Today, of course, this disconnection is not absolute, but it is still usually the case that those who are formally educated achieved their leadership roles through advanced communication skills, while those who exercise leadership in the desert hinterlands achieved their status through intergenerational patterns of descent and election to office. The upper echelons of tribal leadership are highly skilled in fighting in the desert but lack the education or experience necessary for such vital things as policy making, communication through new media, diplomacy, and engagement with the ebb and flow of geopolitics.

By contrast, those whose leadership resides in nongovernmental organizations and agencies of global governance have usually achieved their status from skills of literacy and communication technology acquired through formal education. An example can be seen in the background of Mariam Aboubakrine, current chair of the Permanent Forum on Indigenous Issues, with an MD from the University of Tizi-Ouzou (Algeria) and a master's in humanitarian action from the University of Geneva.[21] She is a prolific user of Facebook, with her page serving as a regular source of updates about the activities of Permanent Forum and the Tuareg-based women's rights organization, Tin Hinan, of which she is a founding member.

The practical effect of this difference is that human rights lobbying has relatively little effect on the decisions taken by those who wield power. Those who make decisions of war and peace, who make or refuse to make alliances with religious extremists and terrorist organizations, are not taking counsel from talented individuals with connections to the institutionalized values of rights- and peace-oriented diplomacy. NGO representation, using the language of human rights and the tools and technologies of formal education, is distant from the practical exercise of political power, which scorns the ambitions of the state and remains deeply embedded in patterns of violence based on desert mobility.

In the civil war of 2012–2013, these formally educated representatives were rendered invisible, with NGO activism and the forced displacement of noncombatants hidden behind the mainstream media's fascination with violence and its perpetrators. This conflict perfectly illustrates some of the problematic dynamics of public attention that I discussed in chapter 4. As I now intend to show, however, it is not only the structures of media, like page-rank algorithms, that consign rights claimants to the furthest margins of public attention. The morally complex situations of violence in which they find themselves also stand in the way of attention and remedy, situations that are almost impossible to explain to anyone who isn't living in them.

The Media Fight for Azawad, 2012–2013

Since the beginning of African peoples' involvement in indigenous human rights forums starting in the mid-1990s, Tuareg representatives have been active participants, recognized by others in the African caucus as assertive and knowledgeable, making their presence felt in such meetings as those of the Working Group on Indigenous Populations, the World Health Organization, and the Permanent Forum on Indigenous Issues and taking a prominent role in Indigenous Peoples of Africa Coordinating Committee (IPACC), a coalition of African Indigenous Peoples' Organizations (IPOs).

Then the world witnessed the occupation of northern Mali in 2012 and 2013 by a loose coalition of Tuareg groups pursuing regional autonomy through the Mouvement National pour la Libération de l'Azawad (MNLA), whose efforts became associated with elements of Al-Qaida au Maghreb Islamique (AQMI) and the rejection by some tribal groups of negotiated peace. This was a very different strategy toward recognition of rights and regional autonomy, one readily associated with violence, strident intolerance, and, to the world's horror, the pillaging of the famous libraries of Timbuktu. The Tuareg groups that briefly allied themselves with Al-Qaida in 2012 did eventually put themselves at arm's length from the most egregious offenses against rights and reason that took place in northern Mali, with the moderates of the MNLA eventually switching sides and fighting alongside the French. But their initial participation in an armed coalition aimed at occupying the north of the country in a bid for secession was sufficient for world opinion—as shaped mainly by major media outlets—to associate *all* the Tuareg tribes and organizations with events that, taken together, can be variously described as a tribal insurrection, a terrorist uprising, or a civil war.

The uprising in which the MNLA declared an independent state on 6 April 2012 and which was mostly destroyed by French and Malian forces early in 2013 was in several key respects fundamentally different from the previous Tuareg insurgencies. For one thing, this time the Tuareg leadership was more ideologically divided, with a secular independence movement first acting alongside and then betrayed by the jihad-oriented Ansar ad-Dīn, which in turn had cultivated relationships with AQMI. Even fragile and shifting alliances in this vexed terrain of ideology and violence produced convictions by association that could not be tolerated by any wider public; laying claim to an independent territory under these circumstances was guaranteed from the outset to fail in its bid for international recognition.

Another key difference is that this time around the struggle for an independent Tuareg homeland in the northern Sahel and central Sahara was not engaged in only by hardened warriors encamped in the remote mountains of the desert using the weapons of war; it was also a battle for the opinion of those who watched the fight from the sidelines: the media-consuming public. Without support from this unknown audience, the hope that a robust regime of autonomy would be offered as the outcome of even a well-prepared and successfully executed war would never be realized. And one of the pivotal devices for leveraging opinion was to make human rights claims based on the status of the Tuareg as an indigenous people whose rights had been consistently violated in a genocidal pattern of displacement and domination. What is more, the international community of states, those "hegemonic actors" whose opinion also matters greatly in any bid for recognized statehood as an outcome of secession, has been very reluctant to give support to any kind of postcolonial claims of self-determination.[22] Here too, the potential for a stable negotiated peace was ultimately rooted in an expectation that appeals to public sympathy could be mobilized by major media outlets.

Despite their political isolation, youth leaders had their own way of pursuing the struggle for the autonomy. It was a group of Tuareg students who organized the November 2010 International Congress of the Youth of the Sahara and, fed up with the broken promises and violations of the peace accords of the 1990s, founded the Mouvement National de l'Azawad (MNA) in a meeting held in Timbuktu. At the conclusion of this meeting, two of the new organization's leaders, Acharatoumane and Boubacar Ag Fadil, were arrested by the Malian police and held in prison on the grounds that they had engaged in treasonous activities damaging to the territorial integrity of the state. They soon became Facebook heroes, appealing to the youth of the Tuareg diaspora and their many sympathizers.[23]

The follow-up to their arrest illustrates some of the power of social media in justice lobbying. Their release was apparently secured (the Malian government did not reveal the reasons) as an outcome of numerous demonstrations and petitions, following which they managed to publish open letters to the would-be citizens of Azawad, the people of Mali, and the people and organizations of the international community.[24] The MNA's lobbying activity included an active Facebook forum as well as an online newspaper, the *Toumast Press*, featuring regular contributions in French and English, each closely argued, well-written, and illustrated profusely with news banners, pop-ups, and images. Before the outbreak of hostilities, the articles on the *Toumast Press* website included a surprisingly well-crafted blog post entitled "Azawad, it's now or never," which described the encouraging geopolitical climate for the Azawad cause that existed at the end of 2011, citing the independence of South Sudan and Eritrea as recent examples of solutions to mistakes made at the time of decolonization.[25]

It was also the technologically imaginative youth who acted on their idea of a television station based in an outpost of the Libyan desert, able to reach across the whole expanse of North Africa from Egypt to Morocco, with some 30 million regular viewers (according to one the proud activists who helped initiate the project). The content of this station included scenes from life in expatriate communities in Europe (mostly France), interviews with people working for human rights, and a scrolling "news ticker" on the bottom of the screen with announcements and summaries of current events. The cost of the station and its transmission, around $70,000 per month, was paid for by Tuaregs who had salaried labor and signed on to make regular contributions of $100 per month. It was robust in its simplicity. When the Libyan government tried to shut it down by force, the people at the station got advance word of the approaching army convoy and were able to get away with their lives and the most essential equipment intact. As smoke rose from the destroyed facilities, the television station continued to broadcast from a laptop on the back of a camel crossing the desert. What ultimately shut it down was Libya's collapse in 2011, when the infrastructure of the country broke down, people lost their jobs, and few were able to contribute their monthly payments.

Despite the technical ingenuity and persistence of the Tuareg activists, however, the publicity they put out during the 2012–2013 war in northern Mali was no match for the state and its access to major media outlets. As the crisis in the north deepened, the Malian press released story after story replete with images of turbaned men with weapons, accompanied by superficially explanatory labels, referring to the Tuareg rebels as "drug traffickers," "armed bandits," and worse, "ex-Gaddafi mercenaries" and "Al-Qaida collaborators." The media war entered the era of "post-truth" and "alternative facts" before the terminology became current. The most damaging thing about these stories is that they contained some measure of truth, not in a way that informed anyone about the ideals and circumstances of the refugees and human rights activists but that tarred everyone with the same brush. The cultivation of opinion through mainstream media excluded the peacemakers and all but precluded a negotiated solution to the conflict.

The focus soon shifted to forms of violence that readily capture headlines and readerships, but in this case at least, not entirely without reason. International news agencies such as the Agence France Press and Bloomberg News readily picked up on stories of atrocities occurring under the Tuareg occupation, including one particularly poignant story concerning 211 cases of sexual violence reported by the UN Special Representative on Sexual Violence in Conflict, "including gang rape, sexual slavery, forced marriages and torture" committed during Tuareg-controlled operations in northern Mali.[26] In this case, the "legacy" media outlets were not weak or absent (as per the view I discuss in chapter 4), but exerted major influence on the perception of events in the central Sahara, to the point of erasing from view the human rights-oriented activists whose claims pointed to state responsibility for mass atrocity.

In an indication that the northern occupation really was out of control, the reporting of crimes against women was not restricted to major media outlets but even included the advocacy-oriented publication, the *Indigenous World 2013*, which reported disturbing information about the Tuareg occupation of the major administrative centers in northern Mali and the imposition of a Salafi version of sharia law on all territories that they occupied: "In Kidal, there were a number of protest marches by Tuareg women, but in vain. For the first time in history, Tuareg men lashed Tuareg women with whips in order to disband the marches."[27] The MNLA did what it could to portray its revolutionary independence movement as not just a movement of Tuareg liberation but one that would benefit all peoples of northern Mali, including the sedentary Songhay and Fulani. Such efforts, however, could go nowhere in the absence of effective governance, and the movement lacked the capacity to build administrative structures, guarantee order, administer justice, or even at a basic level protect the inhabitants under their control from violence.[28] What is worse, by September 2012, the MNLA forces had been attacked by a group of Al-Qaida affiliated jihadists, the Mouvement pour l'Unité et le Jihad en Afrique de l'Ouest, and ousted from Gao, the center of the rebel-occupied region and the declared capital of Azawad. Ironically, in media reporting abroad, despite the actual complex configurations of political violence in the occupied regions, the MNLA also fell victim to an association in the minds of news consumers with the Ansar ad-Dīn and its efforts to violently impose a severe enactment of sharia law in the towns. The Ansar ad-Dīn, in turn, fell victim to the connection between its agenda of religiously inspired regional autonomy and the interests and values of Al-Qaida au Maghreb Islamique. The media—in Mali, France, and internationally—were eager to report on these untidy shifts of alliance and the sordid political values on which they were based.

If this were not bad enough, the media climate was further complicated by "false flag" conspiracy theories directed toward activities of the American CIA and the Algerian secret service, the Département du Renseignement et de la Sécurité (Department of information and security, or DRS), accusing them of being the real cause of jihadist activities in the central Sahara, as a way of discrediting and dismantling the jihadist movement. The information sources for this "false flag" scenario included local, mostly Tuareg military commanders, who were "postulating as to whether Algeria's DRS and its Western allies have been using the Azawad situation to encourage the concentration of 'salafist-jihadists' into the region . . . before 'eradicating' them."[29] As an act of persuasion, this kind of speculation based on third-party "postulation" is indistinguishable from the online imaginings thrown into public discourse more frequently by the alt-right in the United States and Europe, constructed around the motifs of "somebody heard that somebody said," "it could be true," and "it might happen." While the direction taken by this conspiracy theory is sympathetic to the more moderate Tuaregs, it ultimately compromises their efforts toward rights and recognition by throwing *everything*—even the most plausible accounts of shady dealings by secret services—into doubt. This is an example of the new media ecosystem's erosion of journalistic integrity. We are a far cry from the unimpeachable dossiers assembled by Serge Klarsfeld in his pursuit of Nazi war criminals.

Some of the most solid evidence we have of the reality of Al-Qaida's presence in the central Sahara comes from those who escaped it. A Malian member of the Tuareg diaspora, now working for security services in the Gare de Lyon in Paris, told me his story of returning home to Gao in 2011 after the fall of Gaddafi's regime in Libya, where he had received military training and served in the police. He found himself without employment, caught between Al-Qaida and its affiliated drug smugglers on the one hand and the ineffectual Malian army on the other. He chose a rubber raft and the perilous crossing to Italy.

The Human Rights Challenge

One of the strategies used by the Tuaregs to achieve recognition of their will and their rights toward autonomy involved rejecting violence while actively engaging in the international movement of indigenous peoples. This was the context in which I maintained contact with their representatives over the years. The most significant unique characteristic of this movement can be found in its transnational dynamics, not just in lobbying efforts but in the almost global audience that is the target of persuasion and consumption of culture. To make the best use possible of the power of shame, to impose reputational costs on those states and other powers that are the sources of injustice, rights claimants must be heard by wide audiences. This applies especially to indigenous peoples in Africa who have to make claims over and against the rhetoric of centralizing postcolonial states, which frequently attempt to construct a coherent citizenship by refusing recognition of tribal, ethnic, or indigenous difference.

Justice lobbying readily fails in circumstances in which the actions of the state don't attract the attention and indignation of publics, often because of rival interests that stand in the way of popular concern and engagement. With this point in mind, the central challenge faced by Tuareg human rights activists was to generate public sympathy, indignation, and censure toward a Malian state's conduct toward noncombatant Tuaregs and contempt for negotiated settlements that include regimes of regional autonomy. In other circumstances, the Malian state's actions would have provoked outrage, but opposition went nowhere in circumstances of intratribal discord and the inability of any central authority to control either jihadist or secessionist violence.

Under these circumstances, it is telling that Tuareg spokespeople attending the May 2012 meeting of the Permanent Forum on Indigenous Issues in New York didn't touch on the reality of the war and occupation that was taking place in their hoped-for independent territory of Azawad. Aboubacar Albachir, representing TUNFA, a Niger-based Tuareg human rights organization, didn't even mention the occupation of northern Mali in his intervention but chose instead to focus on the history of political usurpation—the "rigidified . . . borders inherited from colonization"—leading up to his immediate concern: the threat to pastoral activity from the licensing of research and exploitation of uranium in the Irhazer Plain.[30]

Saoudata Aboubacrine, representing the organization Tin Hinan (pastoral women) based in Burkina Faso, approached the matter more directly. She pointed out that the constitutional congress of Azawad, held in Gao from the 25th to 27th of April 2012, was met by the Malian government with violent repression, including extrajudicial killings and imprisonments, causing more than 200,000 people to flee the region to the neighboring countries of Algeria, Burkina Faso, Mauritania, and Niger. She asserted the right to self-determination of those Tuaregs living in refugee camps. She then concluded with an oblique reference to the active presence of Al-Qaida–affiliated organizations in northern Mali by stating that the Kel Tamacheq, the original Tuaregs of Azawad, "condemn all action by terrorist groups on their territory, whether making religious claims or not."[31] It is noteworthy that these two Tuareg indigenous representatives each avoided the complex reality of the secessionist crisis, in one instance with single-minded attention to the nomads' environmental interests in northern Niger and in another instance with a narrow focus on Mali's state-sanctioned violence and the plight of refugees.

From this point onward, the silence among the Tuaregs' human rights representatives temporarily deepened. They were absent from the meeting of the Permanent Forum that took place in 2013, just a few months after French forces regained control of northern Mali, even though a half day of this meeting was dedicated to African indigenous peoples' concerns and interests. We can speculate that the complexity of the conflict in northern Mali, the urgency of the ensuing humanitarian crisis, and the experience of losing not just the battle for Azawad but also the battle of opinion effectively silenced those who represented the Tuaregs as collective claimants of human rights.

In this vacuum of reliable reporting and political accountability, the alternating pattern of unfulfilled peace accords and warfare then repeated itself, with the level and reach of violence escalating in unprecedented ways. An agreement between the government of Mali and an alliance of rebel groups known as the Coordination des Mouvements de l'Azawad (CMA) was concluded on 20 June 2015. The global media announced the Agreement for Peace and Reconciliation in Mali, or the Platform of Algiers, with varying degrees of caution, referring to it as "a welcome development" (*The Guardian*) and a "landmark peace deal" (*Al-Jazeera*). The agreement also brought into being the International Committee of Monitoring and Assessment of the Peace, chaired by Algeria, to ensure that the peace was durable. The Tuareg participants in the agreement expressed particular hope that it would finally achieve a regime of regional autonomy for the central Saharan region, to be financed by the international community.

As it turned out, this was false hope. Not long after the signing of the agreement, violence and insecurity returned to northern Mali. In a communiqué from the meeting of its Peace and Security Council on 18 September 2015, the African Union noted "with concern" several "violent incidents" that took place in Anefis in the region of Kidal led by factions opposing the agreement and denying the legitimacy of the CMA. Further attacks across northern Mali produced more refugees and interrupted arrangements for the return of those who had been displaced by previous violence. Shortage of water, acute malnutrition, and an expansion of violence to the central and southern parts of the country were met with only partial and inadequate international support.[32]

Against this backdrop of broken peace, Tuareg rights activists eventually resumed their campaign for regional autonomy as a pathway to justice, with web-based lobbying from servers based in Europe (to prevent their being dismantled by unsympathetic North African governments) and resuming their attendance at international meetings. Despite all their efforts and the poignancy of their words, there was a sense that they were shouting into the wind. People are not moved or mobilized when a state merely refuses to honor its treaty obligations in what is seen as a context of terrorist violence. In these circumstances, the governments responsible for peace can abandon it in bad faith without meaningful censure or cost. The situation in the central Sahara thus returned to something familiar: peace negotiations made impossible by state intransigence on the one hand and an influx of jihadist extremists on the other.

Mariam Wallet Aboubakrine, during a visit to the Law Faculty at McGill University in Montreal, expressed her disappointment to me this way: "The government listens only to the language of arms." Her words came dangerously close to the common rhetoric of revolutionaries: those seeking justice are ignored and subjected to oppression until they act on their grievances with violence. They have no choice. But seen from another angle, her remark was simply a pointed characterization of the reality faced by those pursuing human rights activism and advocacy. How did representatives of a people making use of the instruments and pathways of human rights reach a point of such desperation? The answer follows from a more general question concerning the limits of asserting claims of collective human rights: how, under circumstances in which there is no measure of the reliability of information, does one publicly assert justice claims, particularly when the abuses one suffers are far from the public gaze?

The recent history of factionalism and violence among the pastoralist nomads of the central Sahara demonstrates how difficult it can be to make convincing human rights claims in response to the very circumstances—the forced displacement of peoples and state-sanctioned violence toward its own citizens—that the human rights system was intended to address. The ways that Tuareg activists were able to assemble their resources, build their presence in online forums, and publicly represent the distinctiveness of their people and their rights to autonomy were severely limited by social and political divisions within their own society and the persistent stereotypes in the public consumption of their identity. Images of desert warfare—the heavily armed men with turbans, camels, and Toyota Hiluxes—came to dominate popular understandings of a complex situation. Certainly these images overwhelmed the purely abstract descriptions of the state's unmet rights obligations and hope for future betterment. Simply put, those making use of NGO representation through the language tools and media technologies of formal education were unable to effectively engage with public opinion and encourage popular activism on their behalf. They had little or no influence on rebel leaders whose sources of political power have greater historical depth, who scorn the language of the state and remain deeply embedded in desert life. Nor did they have much influence on the media climate that jingoistically celebrated the defeat of terrorists without acknowledging the human rights claimants and those displaced from their homes—and never even simply took the trouble to distinguish who stood for what.

CHAPTER 6

The Politics of Memory

The historian knows how vulnerable is the whole texture of facts in which we spend our daily life; it is always in danger of being perforated by single lies or torn to shreds by the organized lying of groups, nations, or classes, or denied and distorted, often carefully covered up by reams of falsehoods or simply allowed to fall into oblivion. Facts need testimony to be remembered and trustworthy witnesses to be established in order to find a secure dwelling place in the domain of human affairs.

Hannah Arendt, *Crises of the Republic*

Facebook and Its Counter-Monuments

In a quiet corner of the military cemetery at the Columbiadamm in the Neukölln district of Berlin, there is a memorial to the victims of colonial rule in German Southwest Africa in polished black granite in the shape of the borders of the former colony and of present-day Namibia. It was dedicated on 2 October 2009, 105 years to the day after the infamous "elimination order" was delivered by Lothar von Trotha, the governor of German Southwest Africa and commander of the German Schutztruppen who put down the Ovaherero[1] and Nama rebellions. The ensuing warfare was conducted with a viciousness that took Germany's colonial violence to a new level, resulting in (according to an emerging near-consensus) some 65,000 Ovaherero and 25,000 Nama killed.[2] The granite plaque commemorating this event had been laid, seemingly as a form of historical moral corrective, right next to the "Africa stone," a memorial dedicated in 1907 to some of the German troops who died a "hero's death" (*Heldentod*) in the war.[3]

Curiously, although there was a map at the entrance of the cemetery indicating the graves of famous individuals, there was nothing similar to indicate the location of this memorial to Germany's colonial past. When I asked them about it, the groundskeepers knew nothing about it and could not give directions. I found it a half-hour later by systematically pacing out the length and breadth of the cemetery, looking for the telltale Namibia-shaped plaque I had seen online, and finally found it in the last place left to look, in a shaded corner of the cemetery. With its white lettering filling in and disappearing with dirt, this seemingly neglected memorial plaque seemed to symbolize Germany's ambiguous attitude to the memory and commemoration of what is now recognized by historians as the first mass ethnic killing of the twentieth century—in nonjuridical terms (the Genocide Convention entered into force in 1951) the first genocide of a unprecedentedly violent century that was to have many others. With some trouble, looking closely and mentally filling in the illegible letters, I read the inscription:

> To commemorate the victims of German colonial rule in Namibia 1884–1915, especially of the colonial war of 1904–1907. The District Council and the District Office of Neukölln, Berlin. "Only those who know the past have a future" —Wilhelm von Humboldt.[4]

Humboldt's words resonate here, even as they pose a problem: how does one even begin to "know" the past? How can one ever be *familiar with* or *understand* genocide without witnessing or suffering it?[5]

In keeping with this difficulty, popular opinion in Germany has been largely impervious, if not hostile, to Ovaherero claims for reparations, with the possible exception of calls for the repatriation of skeletal remains of victims of genocide from German museums. The history of mass violence in Germany's colonial past, Reinhardt Kössler notes, "has for long remained thoroughly expunged from national memory."[6] Historical forgetting in Germany is not limited to the prohibition of the political symbols of past authoritarian regimes, but extends to an amnesia that reaches back to the colonial period. Everything that relates to the practices of total destruction in imperial Germany, including the genocide in Namibia and the colonial warfare in Tanzania, have largely been expunged from national memory.[7] The selection and valuation of key episodes in German history as the reference points of memorialization and contrition have their counterpoint in the oblivion of other crimes of the state, more distant in time and space.

To overcome such erasure and to make their presence felt in these challenging conditions, activists have engaged in campaigns of online organization and strategic public visibility, making specific issues the key leverage points of public sympathy. A postcolonial liberation movement has taken form in Germany in recent decades, which represents the cultivation of civil society both in opposition to and *through* the state apparatus. The German government, in its efforts to combat racial intolerance and right-wing extremism (most recently manifested in the rise of the right-wing AfD, or Alternative for Germany, party), contributed to the creation of the very activist network that has been persistently challenging government policy in its promotion of recognition, apology, and reparations for the atrocities that took place under German colonial rule. The Federal Ministry of Economic Cooperation and Development, for example, financed NGO initiatives through a grants program. The most significant of these initiatives took form in May 2000 under the ministries of the interior and justice: the establishment of the Alliance for Democracy and Tolerance against Extremism and Violence (Bündnis für Demokratie und Toleranz gegen Extremismus und Gewalt [BFDT]), which has as its central goal to "make civil society's commitment to democracy and tolerance visible." This is an agenda that explicitly targets right-wing extremism but that has also had the effects of facilitating and emboldening postcolonial activism in its myriad forms.[8]

Among the initiatives partially sponsored by the BFDT is Afrotak TV cyberNomads, an organization that bills itself as "your black multimedia partner for Africa in Germany." This organization is oriented above all toward applying black women's use of media technologies to NGOs that seek to democratically shape the political landscape in Germany. The work of these cyberNomads has included a series of "interventions" assembled online under the heading (in English) "No Amnesty on Genocide Deutschland,"[9] a "decolonial journey through the state structures of Germany," which introduced "into public space the first modern genocide in today's Namibia (1904–1908) by Germany, which, as a predecessor of the Nazi Holocaust in Germany, has played an important role in dealings with 'the other.'"[10] This is one among many examples of the use of media to shape public opinion toward acceptance of postcolonial reform agendas in Germany.

This chapter offers a study of the ways that digital and offline media can be used simultaneously, in this case to create a cumulative reshaping of public ideas about history, memory, guilt, and atonement. The human rights causes of the Ovaherero and Nama, which turn on the history of colonial genocide, have prompted reflection and reform on a variety of issues that have to do with national commemoration and identity in both Namibia (the origin of the genocide victims) and Germany (the source of colonial perpetration).[11] The genocide descendants' campaigns relied on the tried and true strategies of interrupting the flow of acceptable speech, using cultivated moments of confusion, curiosity, horror, and amusement to plant messages that run counter to dominant national discourses.

One of the things that makes these efforts noteworthy is their omnivorous use of media. Nothing was too technologically simple or sophisticated to make it inappropriate for use, including distributing leaflets, staging theatre, installing "counter-monuments," reaching out to the mainstream press, and, of course, cultivating sympathy and common cause through regularly curated Facebook pages, all oriented toward challenging state efforts to command attention and secure the legitimacy of power, publicly displaying artifacts of oppression.

Another feature of the Ovaherero-Nama case that stands out is the way that their expression of grievances through a transnational multimedia campaign contributed to their long history of identity formation in the aftermath of genocidal violence. Through their campaign for justice, they have created clearer symbols and structures of belonging. In the terms used by British social anthropologist M. G. Smith, the Ovaherero and Nama, in the long aftermath of genocide, have been able to reshape themselves into "corporate groups" with determinate boundaries and memberships, "defined by common disabilities and burdens."[12] Bringing this observation into the present, they offer a clear illustration of the use of the internet lobbying to simultaneously leverage appeals to public audiences and cultivate the symbols and strategies of collective belonging.[13]

Understanding the full range of efforts to represent contested national histories takes us once again from one of the oldest technologies of communication, the monument, to one of the newest: Facebook. Hierarchical societies have had an affinity for monuments for as long as they have been in existence. In their monumentality, they represent the power of the state to assemble resources that go far beyond those of any subject, in a way intended to evoke awe and, ultimately, obedience. In the symbols they depict, they put state versions of history and legitimate rule into permanent form, with their materials and size of assemblage conveying ideas of permanence and the solidity of existing forms of power. In honoring fallen soldiers of the colonial era with their names inscribed on a boulder large enough to be immovable by unaided human power, the state-sponsored promotion of the legitimacy of the sacrifice of those who gave loyal service to colonial power is made manifest in a way that cannot be easily contested.

That is, not until the popular media arose that gave voice to insurgent ideas and loyalties. The counter-monument of the Columbiadamm cemetery is but one in a series of efforts oriented toward undoing the received ideas of Germany's colonial past. As its accumulations of dirt and leaves reveal, the physical counter-monument of the activist campaign, once erected and publicly dedicated with all the requisite solemnity and speeches, took on a virtual life in images posted online.[14] Digital space is where dissenting versions of history are given greater reach, where they create community and situate their moral authority. Historical memories and the monuments representing them are digitally rendered and contested online, sometimes leaving their more tangible forms behind, subjected to the invasions of mold and humus.

One of the first steps in the Ovaherero-Nama campaign was digitalization of the photographic archive: black and white images taken by German soldiers of German soldiers posing next to black bodies hanging from trees, of naked Ovaherero women standing in a row, and of naked adolescent girls (providing material proof of their sexual exploitation), of decapitated heads preserved for collection (photographed face-forward and in profile, like a mug shot), of Ovaherero boys in chains standing next to their colonial masters, and a postcard image of German soldiers packing skulls in shipping crates. These all have become the evidentiary bedrock of the Ovaherero and Nama claims to justice. As with all monuments, from the images alone the viewer cannot see the *extent* of the horror, the numbers of lives destroyed and taken, but in their subject matter, in what they are meant to reveal, they convey something more readily recognizable: individual *moments* of horror and the unmistakable signs of wider campaigns of genocidal violence. In doing so, they evoke an imaginary of the testimony that *might have been* narrated by living survivors.

Consistent with these digital archival sources, Facebook has become the social media site of choice for Ovaherero and Nama activism. The platform's ability to make connections between images and social worlds lent itself to transnational activist networking. Facebook has in this case become a site of vernacularization, the process by which intermediaries such as NGO and social movement activists interpret the cultural world of transnational modernity for local claimants.[15] Unlike the usual top-down scenarios of human rights cultural brokerage, however, there is more room on Facebook for "upstreaming," for the reversal of the flow of ideas and representations of identity, originating with the subjects of human rights claims.[16]

The images posted as headers in Facebook groups have qualities that are monumental or "counter-monumental." There is, for one thing, a kind of permanence to them—true, not the heavy three-dimensional permanence of state monuments constructed from stone or bronze, but certainly greater permanence than the usual social media postings that get overlaid by new information and provocation. Seen again and again as group members visit the site, the images invite deeper reflection than those meant to be seen once and scrolled past, much like physical monuments erected in public space that people encounter repeatedly and usually without reflection in the course of their daily routines. There is an infiltration into consciousness (or deeper) that happens through repetition of an image or idea. Facebook Group pages engage in this repetition as they represent a community, or, in the case of those groups set up in support of a claim or a cause, a polity. Like monuments, the images their leaders or curators select to represent the group on Facebook define the core values of citizenship. And in the case of the Ovaherero, that citizenship and sense of belonging to a community is defined by the image and idea of genocide. (The Facebook algorithm, meanwhile, occasionally sends up invitations to join the Ovaherero wedding group, with over 4,000 members. It depicts a smiling woman in her white finery, not registering the uncomfortable juxtaposition of mass killing [figure 6] with the joining of happy couples.)

There is no substitute for touching the primal core of human emotions, and the digital archive with its raw imagery gave an element of irrefutability to the claim of genocide. Whereas the Ovaherero and Nama did not succeed in their efforts to secure reparations from the German government and corporations for the 1904–1908 genocide, they did manage to secure recognition of the genocide (using the term), of the fact that it had occurred and was a mass crime even in

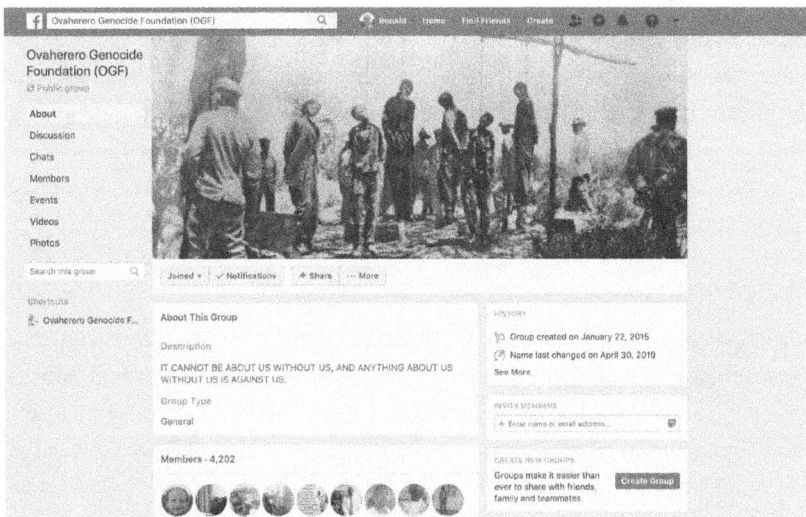

Figure 6 A screen capture of the Ovaherero Genocide Foundation's Facebook page.

the era of colonial domination. But even here, public opinion in Germany was divided. Many did not want their national identity as a leader of the Enlightenment to be, as they saw it, sullied by a reframing of national glories as colonial crimes. They did not want, as one speaker at a museum forum in Leipzig put it, the "shame" of the atrocities committed in the colonial era applied to the German people as a whole.[17] Others more quietly saw the recognition and clear-eyed examination of the crimes of the state as a positive virtue, a source of distinct German identity and national pride.[18]

For the Namibian activists and their allies, however, leveraging this point of contested opinion was not easy. There were no living witnesses to bring the victim experience into the present. The violence could be conveyed only partly by photographs and stylized with greater permanence in the Columbiadamm memorial, but there was little to provoke the deeper, more visceral reaches of fascination, horror, and nurturing, sympathetic sorrow that come from witness testimony.

There was one corner of their protest, however, where the Ovaherero and Nama did strike a chord of public compassion and institutional embarrassment. They were able to galvanize attention and action around the fact that human remains from the genocide were being stored in German museums and research facilities. The removal and sequestration of human remains is a violation of the universal norms (even instincts) of respect and ritualization surrounding death. The dead speak for (and to) the Ovaherero and Nama from the drawers of the museums and research institutions where they are held. And they speak to the public, their image reaching the imaginations and indignations of those who feel the enormity of a mass crime—not so much the crime of historical genocide, which to many Germans can be safely buried in the past, but that of the ongoing separation of survivors from their ancestors and of keeping the dead in a place and condition of desecration.

Recognition

On 10 July 2015 the German government and parliament officially designated the killing of some 65,000 Ovaherero and 25,000 Nama in German Southwest Africa from 1904 to 1908 "genocide" (*Völkermord*).[19] After years of official denial, this came as a surprise to many observers; but perhaps Germany's volte-face was to be expected when seen in the context of active and effective lobbying by (and between) activists in Namibia and Germany.[20] Since 2016 negotiations have taken place between special envoys appointed by the German and Namibian governments, creating a sense of exclusion among Ovaherero and Nama activists.

The issues under discussion in the negotiations, above all the issue of framing compensation as "reparations" for events going back more than a century, remain fraught.[21] Despite Germany's gestures at recognition, postcolonial activists often note the strong resistance that continues to meet their arguments for treating the legacy of the 1904–1908 genocide as a current source of political responsibility. The German government has a tendency to simply ignore them or to deal superficially with their demands.[22] Activists point to an element of contrition-saturation that stops Germany's "memory culture" at the Second World War and refuses to extend it back through intergenerational memory into the murky realm of colonial genocide.[23] Germany's resistance to recognition of its colonial violence has parallels in Namibia's response to the genocide, in which the SWAPO-dominated government has emphasized the independence struggle of the 1970s and 1980s as the central reference point of national identity, while approaching the era of German colonization as "a sad history," which had to be put aside for the nation to look toward its future.[24]

The Ovaherero-Nama genocide had its origin in a colonial war that took place from 1904 to 1908, with violence continuing thereafter with the operation of internment camps—including an infamous labor camp on Shark Island with an estimated 80 percent mortality rate, where inmates were starved, beaten, exposed to the elements, and worked to death.[25] Isabel Hull attributes the mass atrocities committed in German Southwest Africa to an imperial military culture with marked proclivities toward "absolute destruction" in response to political insurrection.[26] The manifestation of this proclivity in the war of annihilation directed against the Ovaherero and Nama precedes the much better-known Armenian genocide by roughly a decade.[27] This has more significance than the symbolic value of temporal precedence as "the first genocide of the twentieth century." It means in addition that it is the first publicly prominent genocide to be claimed without the immediate experiences of victimhood and survival, through acts and contests of memory making and commemoration.

Germany, needless to say, is not alone in its colonial legacy of institutionalized racism, booty capitalism, and mass violence. The suppression of the so-called Mau Mau rebellion in British Kenya between 1952 and 1960, in which the Kikuyu were killed in the tens of thousands, is a clear example of the worst forms of dispossession and violence that took place in European colonies. It is also a present source of reparations claims.[28] To this could be added the violence of

the French in Algeria and elsewhere—the Harkis, survivors of both revenge killings and French apathy that accompanied their lack of support for the collapsed
colony, are one of the most active postcolonial survivor groups today.[29] Arguably
surpassing these examples in terms of the scope of suffering and numbers of dead,
the horrors of the Belgian Congo clearly meet today's criteria for war crimes and
genocide.[30] Yet Germany has gone (or been taken) the furthest among European
states in recognizing the actions of its predecessor regime as genocide, of making
acts of recognition and contrition, and—if legal claims currently being pursued
by the Ovaherero and Nama are successful—making reparations for the worst
crimes of colonial rule.[31]

How did this happen? The Ovaherero and Nama claimants and the advocates
supporting them have had to be imaginative in bringing attention to their cause,
above all because there is an absence of living witnesses to the crimes on which
their claims are based, no one who can directly invoke memories of violence
and its consequences—as Hannah Arendt says in the epigraph to this chapter,
"facts need testimony to be remembered." The efforts to overcome this absence
of witnessing center upon what is sometimes referred to as the "Windhoek to
Auschwitz thesis," the teleological line from the Ovaherero to the Nazi genocide, which has become the source of contested terrains in Namibian, German,
and comparative histories.[32] Through their emphasis on a connection to the
Holocaust, Ovaherero and Nama claimants have taken the concept and claims
of genocide into the new, historically more distant realm of colonial violence.

Germany's "Remembrance Culture"[33]

There is a central difference between the crimes of the Nazi regime and those
of German Southwest Africa: in the post–World War II era Germany was faced
with living survivors of its actions, citizens whose dignity and legal personhood
urgently had to be restored. Similarly (in this one way), the fall of the Berlin Wall
in 1990 and the process of German reunification produced a sense of a new sense
of kinship (*Verbrüderung*) that crossed the former boundary of the wall,[34] but it
also produced claims of recognition from the survivors of political persecution
under the GDR. In chapter 4 I discussed the observation that the involvement
of victims in justice causes is central to the cultivation of public sympathy and
ultimately the success of the rights campaign.[35] Consistent with this, there was
in German public culture a distinct presence of victims of the crimes of the state
who eventually came to refer to themselves as "survivors." More broadly, this
presence contributed to a popular sensibility that recognized the mass suffering

caused by political crime and that demanded reparations.[36] In Namibia, by contrast, memory did not come from direct witnesses and survivors of violence but was intergenerational, passed on in oral tradition. Apart from some gestures from the Rhenish missionaries, who even in the midst of mass killing observed that solace for the disastrous consequences of the "unhappy insurrection" was to be found in the "protection of the German Christian people,"[37] recognition and restoration in the postwar period were all but absent, coming instead mainly from within the shattered communities of survivors themselves.

One of the distinct features of German public space is the extent to which it is dedicated to remembrance, not in an honor-our-fallen-heroes sort of way but in a consistently direct acknowledgment of the legacies of despotism, war, and genocidal violence.[38] In Germany, Ovaherero and Nama genocide claimants encountered a state and a public that was already accustomed to acknowledging the crimes of earlier regimes. Namibian claimants could point to the 1904–1908 genocide as an early manifestation of the same impulse toward mass killing that later took form in the Holocaust. Their challenge has been to persuade the German public that, despite the distance in time of colonial violence, acknowledging and paying reparations for the mass crimes of the colonial regime of German Southwest Africa could and should take place in similar ways to the state's response to the aftermath of World War II and the Holocaust.

The main areas of tourist traffic in Berlin, particularly near the Reichstag and the Brandenburg Gate, are the central loci of state-sanctioned remembrance.[39] The centerpiece of Germany's Holocaust remembrance, the Memorial to the Murdered Jews of Europe (Denkmal für die ermordeten Juden Europas), is situated near what used to be center of the Nazi administration, including its chancellery building and Hitler's bunker (now a parking lot). Here, the symbols of the defeated and rejected regimes of Germany's past have been systematically removed and criminalized, then overlaid with commemorations of the mass murder they perpetrated, intended, according to a Bundestag decision, to keep alive "the memory of an inconceivable incident in German history" which can "serve as a reminder to all future generations."[40]

The criminalization of the symbols of National Socialism is a side of Germany's "memory culture" (Gedächtniskultur) that is not often emphasized. It does occasionally emerge, however, in the context of a particularly prominent example of its breach, as when, in August 2017, the police detained two Chinese tourists after they were caught photographing themselves making Nazi salutes in front of the Reichstag.[41] Memorialization in Germany is premised on the prohibition of the political use of those symbols or propaganda that invoke "constitutionally prohibited" (*verfassungswidriger*) organizations, punishable by up to three years in prison. Attention to the mass crimes of historical enemies extends toward circumscribing their symbols and preventing their return in new form. This prohibition has become especially relevant in the context of a resurgent far right.

As the national capital not only of the Federal Republic of Germany but also of two major criminal regimes of the twentieth century, Berlin is arguably the epicenter of the culture of memory in Germany. It is difficult to walk anywhere in this city without encountering purposeful and often powerfully evocative efforts to memorialize a dark past that is still within living memory. The acknowledgment of war, destruction, unjust laws, and state-sanctioned killing and genocide are unmistakably and unavoidably on public display. Situating oneself in any area of concentration by tourists will bring memorialization to the senses. Take, for example, the area outside the famous KaDeWe, a luxury department store that under the Nazis was confiscated from its Jewish owners and used as a gathering point of high-ranking officers. Across the street from the store, outside the Wittenburgplatz U-Bahn station, a large, unadorned panel presents a list of Nazi death camps, accompanied by no narrative. Nearby, the full-throated bell of the Kaiser Wilhelm Memorial Church, the famous Gedächtniskirche (memory church) that has been left standing in architecturally reworked ruins, resonates across the entire district of Charlottenburg as a periodic aural reminder of the destructiveness of war.

Looking at the sidewalk, here as in in many other parts of Berlin, one can find examples of everyday remembrance in the *Stolpersteine*—literally "stumbling stones," an allusion to the fact that these small commemorative brass blocks are usually encountered by accident, where one walks. They have been installed in front of the homes or places of work of the victims of National Socialism. Each "stone" marks the Nazi regime's crime against one person, usually under the heading "*Hier Wohnte*" (here lived), followed by their name, date of birth, date of arrest, year of deportation to a concentration camp, and the date of their murder.[42] There is an element of mass participation in the project, with a web page advertising the project, where an online donation of 120 euros will pay for the installation of one *Stolperstein* memorial (http://www.stolpersteine.eu/en/home/). Using this participatory approach, the artist Günter Demning has directed the placement of some 61,000 of these stones in more than 1,200 towns and cities in twenty-two European countries.

Taken together, the regular encounter with this kind of memorialization produces (and reflects) the phenomenon referred to as "memory culture" or "remembrance culture" (*Gedächtniskultur* or *Gedenkkultur*), which I understand as a political ethic of remembrance, based on carefully cultivated forms of perception, classification, and affect. There is a distinct element of habitus at work in Berlin's memorialization, a publicly willed, deep intrusion of historical memory into everyday life. Habermas referred to this ethic of remembrance as "the obligation we in Germany have—even if no one else is prepared to take it upon themselves any longer—to keep alive the memory of the suffering of those murdered at the hands of Germans, and we must keep this memory alive quite openly and not just in our own minds."[43] The design and construction of memorials (*Denkmal-Arbeit*) in this context has been a "tortured, self-reflective, even paralyzing preoccupation" in which "every monument, at every turn, is endlessly scrutinized, explicated, and debated.[44] This ethic of memorialization does not so much take the form of personal regret—most Germans did not participate in the decisions that produced the Nazi or communist regimes—as acknowledgment of the moral choices of the past and an awareness of the ever-present danger of similar forms of tyranny, should these lessons be forgotten.

The power of this message and the constitutional prohibition of certain symbols and political allegiances do not entirely repress dissent. The entry of the far-right Alternative for Germany (AfD) party into parliament in 2017 marks the growing popularity of starkly nationalist ideas, including the rejection of contrition toward Germany's past and construction of war remembrance around honoring the soldiers who faithfully served the country.

Protest also takes the opposite direction, toward resistance to incomplete memorialization. A small group of former political prisoners based in Berlin, for example, is active in its defense of memorialization of the abuses of the GDR regime. They were vociferous in their efforts to retain a memorial to victims of the GDR that took the form of crosses representing the victims of the Berlin Wall and the barbed wire at the intersection of Friedrichstraße and Zimmerstraße, chaining themselves to the crosses in protest before the memorial was removed on 5 July 2004. A web page sponsored by the former prisoners continues this struggle against the material removal of memory with considerable attention to detail. For example, in their website they disapprovingly took note of the removal of the apertures from an arrest cell and the removal of bars from a cell nicknamed the "tiger cage" in the Lindenstraße Memorial for Victims of Political Violence in the Twentieth Century (Gedenkstätte Lindenstraße für die Opfer politischer Gewalt im 20 Jahrhundert).[45]

I met one of these protesters, Gustav Rust, a prisoner for nine years under the Ulbricht and Honecker regimes, sitting on a bench near a row of white memorial crosses not far from the Reichstag in Berlin. Next to the crosses he had attached leaflets (in German, English, French, Spanish, and Italian) and a tin can for donations. On his left wrist he was wearing the handcuffs he once wore in custody, combining the public display of an artifact of historical memory with the embodiment and performance of victimhood. The main focus of his indignation was the SPD/PDS "comrades" (*Genossen*) and Greens who "while they can, are getting rid of a number of memorials to victims of Bolshevik terror," including the white crosses memorializing victims of the Berlin Wall.

This example tells us that the power and means that are brought to bear in the struggle over memorialization might be lopsided, but that does not always stop activists from pressing for public space and artifacts of memory to commemorate their historical causes and represent their experiences of suffering. In doing so, they are not daunted by the superior resources of the state and use whatever technologies of public outreach are at their disposal, including artifacts (crosses and handcuffs), leaflets, and online activism.

Postconflict Constructions of Memory

As long as centralized political power has existed, those who hold power have created official traditions and histories, shaped collective memory of key facts, promoted origin myths, and encouraged ostracism of the vanquished. The human rights era created ways for minorities and representatives of victims to contest official narratives and to make their versions of history heard. In the name of human rights, the crimes of the state became part of new historical narratives.[46] Memory became as much a tool for social reform as it was of state legitimacy.

Activism, online and "real world," has therefore often involved contests over the symbols intended to lend legitimacy to state power. The 1992 quincentenary of Columbus's landing in the Caribbean, for example, was marked by an invocation of the American Holocaust and corresponding protests against monuments dedicated to the explorer.[47] Around the same time, the commemoration of Britain's air force battle against Germany in World War II provoked a vocal reaction by the Peace Pledge Union and by the mayors of Cologne and Dresden against honoring Sir Arthur Harris, the architect of the British saturation raids.[48] More recent, and more directly related to Ovaherero-Nama genocide commemoration, was the South African protest against statues depicting Cecil Rhodes, the zealous promoter of British imperialism in the Cape Colony, in part through a twitter campaign based on the hashtag #RhodesMustFall and its later abbreviation #RMF.[49] The examples could be multiplied to illustrate the connection between retrospective recognition of the crimes of the state and the monuments that take on contested meaning in depicting, and at times celebrating, the symbols and actions of those states.

There are two central features of the Ovaherero-Nama effort to bring colonial violence into the politics of the present that I will emphasize here. First, the lack of living witnesses to the crimes of the colonial state has shifted the emphasis of the Ovaherero-Nama justice campaign from victim narratives of suffering—the usual way to express and establish justice claimants' connections to experiences of mass violence—to symbolic, historical, and heritage-based representation of genocide and political abuse. Transgenerational trauma, centered on the massive harm the ancestors suffered at the hands of their colonial oppressors, becomes a way to cultivate group identity.[50] The material markers of genocide give reality to this trauma. They establish a link across generations in a campaign that substituted heritage for testimony as the central source of connection with potentially sympathetic public audiences. The human remains stolen under reprehensible circumstances during the 1904–1908 genocide and stored in German museums and research institutes acquired significant meaning for the Ovaherero and Nama as "ancestors" who manifested a spiritual connection with the past.

Second, the campaign was transnationalized to a site of greater strategic support, public sympathy, and political resonance. The field of awareness-raising and contestation that had the most potential to further their cause was not in Namibia, where the Ovaherero and Nama lack political influence and public visibility, but in the more fertile ground of Germany, where they have been able to situate themselves within a postcolonial movement that is actively struggling to reshape the ideas and material representations of Germany's colonial history. Comparing the dedication of public space to permanent symbols of genocide in Namibia and Germany thus involves more than an inquiry into starkly different political uses of symbolism and different public sensibilities toward historical memory. We also need to consider the impact of the "travelling model" of key concepts, in this case the claims making oriented toward the concept of genocide, in which ideas, as Richard Rottenburg puts it, "circulate from one social world to another, from one frame of reference to another," and in the process are "adopted, appropriated, and altered."[51] Their reappropriation provides them a new frame of reference and adds energy to their wider communication.

Consistent with this notion of the amplified impact of translated and re-purposed ideas, the contests of history that the Ovaherero and Nama fought in Namibia were adapted to and repurposed for their claims in Germany. There they encountered well-established, materially manifested, state-sanctioned pathways of recognition and contrition that had been applied toward the victims of the brutal tyrannies and mass violence of the Nazi and GDR regimes as a way to deepen the legitimacy of the post-Holocaust and reunified German regimes. In this new context, the Ovaherero claims of colonial genocide gained leverage through—and tested the limits of—Germany's habitus of historical contrition.

When Ovaherero spokesperson Kambanda Veii, in a sparsely attended news conference in Berlin in 2017, interrupted a line of questioning on Namibian politics by saying, "We are fighting against *two* governments." She was attempting to simplify for journalists the complex, transnational dynamics of the Ovaherero genocide claims.[52] Elaborating slightly on her point, a simplified outline of the struggle might be put as follows: the Ovaherero and Nama are bringing their own movement toward postgenocide reconstruction of colonial history (postconflict situation 1) in a strategy of claims making with reference to Germany's Holocaust remembrance (postconflict situation 2) *and* an effort to reshape Namibia's dominant narrative of national identity, which places emphasis on SWAPO's war of independence against South Africa (postconflict situation 3). Like many activist efforts, the Ovaherero's and Nama's genocide-awareness campaign is multisited, multimedia oriented, and involves a complex interplay of interests and identities.

There is a canonical element to the selection of those events and personages that represent collective identity, usually emphasizing the core qualities of identity promoted by the state.[53] To varying degrees and in different ways, state-sponsored representations of history are propagandistic, reducing complex historical events, acts, or periods to simple symbolic representations of the power and perspectives of those who build them and condensing their subject's significance to "a few patriotic lessons frozen for all time."[54] Their intention is often to appeal to the emotions of their audience through stirring, stylized presentations of the past, highlighting the heroism and sacrifice of those who struggled for the foundation of the state or the death and suffering of those who were the victims of its previous regimes.

Within a national community, postconflict commemorative acts and artifacts, including their differences and contestations, can constitute fundamental sources of the self, something that Jürgen Habermas referred to as a "solidarity of the memory," ideally involving "a reflective and keenly scrutinizing attitude towards one's own identity-creating traditions."[55] Those who seek to promote an alternative vision of national identity are challenged to influence opinion, to repudiate rival loyalties and shape public memory, in some cases to persuade public audiences of the need for states to formally recognize and address the history of violence and the festering grievances that remain. Under these circumstances, contests over memorialization readily cross into the realm of "identity politics" focused on the public representation of national history.

Within these postconflict contests, it is important to be attentive to the constitutional dimension to states' choices of flagship monuments, those structures and spaces that depict a central, hegemonic narrative of the state. Those monuments that proudly proclaim victory in war are intended to solidify the legitimacy of conquest and the regimes on which it is founded.[56] Sicard's 1921 Monument à la Convention Nationale in the Panthéon in Paris is a classic example, with its towering figure of Marianne dispassionately holding a sword, flanked by her supplicating generals, producing an explicit reference to the constitutional connection between warfare and nation building.

Then there are the less common commemorative displays like the Holocaust memorials in Germany that mark the rejection of a defeated regime's most misguided and catastrophic policies and that at the same time promote a more global conception of the dignity and worth of the human person. Here too, in the design of the monuments and the choice of poetic words with which they are engraved, there is a clear intention to mark a constitutional turning point. The intention of such memorials is not only to mark the new dispensation of a state regime but also to literally concretize its difference from a previous order. Without monuments, constitutions are words on paper. With them, they take on greater life, legitimacy, and patriotic appeal. Monuments, in other words, can be quintessentially constitutional artifacts, publicly oriented extensions of the arid legal documents that form the foundations of political communities.

The power dynamics behind the politics of memory are almost always one-sided in favor of the state. Michel-Rolf Trouillot observes, "The production of historical narratives involves the uneven contribution of competing groups and individuals who have unequal access to the means for such production."[57] This is particularly true of the expression of history through monuments. Those works of art and engineering that mark the central events of state history are overwhelmingly under the control of the state. Few challengers are able to summon the resources needed, for example, to construct museums or erect monuments worthy of public attention. There are also mechanisms of legal control over the sanctioned use of public space at work in the placement and design of commemorative artifacts. To be effective, the creation of alternative public memory calls for access to resources such as education, specialized technical expertise, computer skills, access to printers and publishing houses, and possibly even the resources to erect a monument or run a private museum. Behind the shared grievances of human rights activism and identity, there are wide differences in the means available to activists to elicit the public recognition and support needed to influence the symbols and behaviour of states. Contestation, when it does manifest itself, usually occurs in conditions of grossly unequal access to the publicly accessible media of commemoration.[58]

Despite their state sanctioned, constitutionally oriented canonization, however, memorials are works in progress, subject to continual construction and negotiation of meaning.[59] Regimes come and go, political alignments and priorities change, and with them the conceptualizations and commemorations of the past become contested and transformed. Names are given, refused, debated, and reassigned to specific events that are marked with street names, tablets, plaques, or monuments, with or without spatial connections to the happenings in question. As works in progress, they are actively questioned, serving as points of reference for claims over history and as strategic sites for the definition and political mobilization of communities.[60] Even in a high-technology world, contested opinion can still galvanize around a form of communication that is nearly as old as art itself.

For as long as states have made use of monuments and other media to secure legitimacy, they have claimed their use of these media for themselves. One response to the resulting inequality in the means of communication is for marginalized groups to engage in their own everyday acts of commemoration with whatever technologies are at their disposal, in ways that directly challenge the dominant narrative of the state. Defacing monuments with controversial meaning is the most common strategy used to draw attention to historical grievance. Hence, the proverbial red paint on monuments intended to disrupt narratives of glory and draw attention to a counternarrative of violence. This is a form of the kind of strategic messaging of graffiti that I discuss in chapter 1—similarly amplified with posted images and discussion threads on social media.

Dissent can take form in ways that replicate the state's power to construct memory—for example, through "counter-monuments" in which new installations offer contested meaning to dominant structures of architecture and memorialization.[61] Such monuments are not always technologically sophisticated but can be manifested in things like personal collections of artifacts and off-the-beaten-track roadside museums. In Namibia, for example, memorial dissent is sometimes practiced by German descendants who maintained collections of colonial artifacts in ramshackle museums, access to which requires asking a nearby shopkeeper for a key.[62]

Although issues of unequal access to the education and technology necessary for marginalized groups to organize and articulate grievances are ever present, use of new information technologies has gone a long way toward overcoming the differential access to the media in which contested visions of violence and memory are stored and by which they are conveyed. Where subaltern groups have access to the necessary expertise and technology, influencing public memory is often initiated online and then manifested in "real world" activism and memorialization. In this sense, Facebook has become a platform for counter-monuments, sites of curated and contested memory.

Postcolonial Justice Campaigns

A campaign given the title Völkermord verjährt nicht (genocide does not age) posted a press release on 13 June 2016, which called on the German government to "immediately recognize the genocide of the Ovaherero and Nama!"[63] and went on to explain to its audience, "The long-standing refusal of successive federal governments to recognize the genocide in Namibia remains a poor testimony to the Federal Republic of Germany and does not fit in any way with their self-image as a 'memory world champion' and the associated critical handling of their own history."[64]

The consistent target of the activism in Germany by and on behalf of the Ovaherero and Nama claimants from Namibia was the moral limit of the Second World War and the Holocaust, beyond which the German public and their politicians seemed unwilling to go. The idea of going further back in time to the attitudes and events of Germany's colonies as sources of recognition, contrition, and remediation for a harm of the state and its people was, and remains, widely resisted. In this context, Germany's Holocaust memorials and its celebrated "memory culture" served as reference points for dissent, for accusations of hypocrisy and appeals for sympathy by collaborative activist organisations and campaigns.

Under the impetus of this government-sponsored activist network, challenging the colonial geography of urban space became an issue that resonated widely. Street names that evoked the rejected values of domination were a particular target of protest.[65] Von Trotha Straße in Munich first went through a change of official signification in 1994, with the name remaining the same but the justification being shifted from the infamous colonial governor who commanded the troops and instigated the massacres in the 1904–1908 war in German Southwest Africa to his distinguished family. (As one Berlin-based activist pointed out to me, this would be like keeping a "Hitler Street" unchanged in the postwar period by making it in honor of Hitler's family or someone else with the same name.) Von Trotha Straße was officially renamed Hererostraße in 2007, though not without resistance from residents of the neighborhood who argued that the Herero, having historically committed atrocities against Bushmen, should not be commemorated.[66]

Activists were not as successful in changing the name of the street and U-bahn station, Mohrenstraße, or Moor Street, in Berlin. For the past ten years or so this name has been a focus of activist attention because of its historical connection to seventeenth-century practices in which Africans from Ghana (then referred to as Moors) were brought to Berlin as troubadours, serving in slavery as the exotic décor of elite households. In 2009 human rights activists, getting nowhere in their efforts to change the name, resorted to changing it unofficially, with a protester dressed in a pink agit-prop rabbit costume ceremoniously climbing a ladder and adding umlauts to the *o* in Mohrenstraße to make it Möhrenstraße, or Carrot Street.[67] This event, with its distinct humor, was duly covered in the national press and in a longer-term campaign, Pink Rabbit Against Germany, oriented toward disrupting Germany's "national nonsense." This campaign had both an online and real-world theatrical presence. The pink-rabbit campaigners introduced their mascot at strategic events to bring attention to the hypocrisy inherent in the absence of memory of German colonial history, framing the construction of state-financed museums as "reactionary architectural projects" and "colonial exhibition fantasies." Its exploits were duly posted online, along with a campaign manifesto.[68]

There are signs that such small individual efforts have a cumulative effect on policy and political discourse. In the case of the Ovaherero-Nama campaign, this began where their activism has its greatest power of persuasion: in the issue of human remains from colonial mass killings in the collections of museums and research institutes in Germany, quite possibly including the unrecovered skull of Cornelius Frederick, the leader of the Nama rebellion. Facebook became the social media site of choice for promoting this cause and forging links with activist networks across the postcolonial spectrum. The organization Berlin Postkolonial, run by activist organizer Christian Kopp, took a leading role in coordinating the Ovaherero-Nama campaign in Germany. The hashtags #genozïdverjährtnicht (genocide does not age) and #setthemfree (a reference to the remains of those killed in the genocide being kept in German museums and research institutions; figure 7) formed online assembly points for information, opinion, and activist organization.

The transmedia activism that has ultimately begun to make inroads into colonial amnesia has been based on a network of small NGO initiatives, often

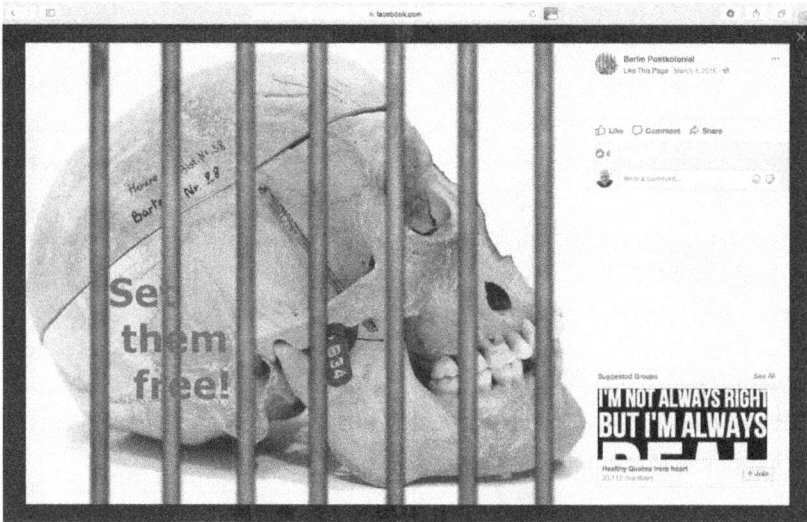

Figure 7 A screen capture of Berlin Postkolonial's Facebook page.

the work of talented and committed individuals leading small groups of collaborators. These organizations then reached out to potential partners through
online initiatives that multiplied and intensified the public impact of their acts
of communication. An online petition, circulated and submitted to the Bundestag in 2012 under the banner "No Amnesty on Genocide!" was sponsored by
a remarkable diversity of web-connected organizations, including AfricAvenir
International, Afrika-Rat Berlin-Brandenburg, Afrika-Rat Nord, AFROTAK
TV cyberNomads, Arbeitskreis Panafrikanismus München (AKPM), Arte-
fakte//anti-humboldt, Berliner Entwicklungspolitischer Ratschlag (BER), Berlin
Postkolonial, Deutsch-Afrikanische Gesellschaft Berlin (DAFRIG), Initiative
Schwarze Menschen in Deutschland (ISD-Bund), and Solidaritätsdienst International (SODI). The petition was billed as an "appeal to the Members of
the German Bundestag for recognition of and compensation for the genocide
in the former colony of 'German South-West Africa.'" It offers a justification
for using the principles of transitional justice and the tools of diplomacy to
overcome the challenges of redressing colonial violence and includes a call
to the Bundestag to promote "a postcolonial remembrance culture" that will
"support the decolonization of public space in Germany." This means, among
other things, putting a stop to "the continued glorification of colonial criminals"
with street names and memorials and instead paying tribute to the key figures
of African resistance.[69]

The Politics of Memory in Namibia

In Namibia, three historical narratives are situated in a framework of represen-
tational competition and interconnection: (1) the state-sanctioned narrative of
armed liberation struggle led by SWAPO, which, in a secondary narrative, has
produced the peaceful racial mosaic that is today's Namibia; (2) the German-
descendant narrative of conquest and survival in a harsh land, with a secondary
narrative of military heroism and genocide denial in historical conceptions of
the colonial wars; and (3) narratives of survivor resilience and ongoing marginalization of the colonized Ovaherero and Nama. The context that brings these
narratives together is the dominance of the state narrative and the competition
between German descendants and self-identifying survivors of the Ovaherero-
Nama genocide over the terms of an alternative.

Monumental Narratives of the Namibian State

The way that a government chooses to promote or participate (or not) in acts of
commemoration usually reflects political fault lines, but often in a way that is
indirect and occluded. Two central narratives of Namibia's public representation
of national history and identity, at least in the first fifteen years after independence in 1990, were the country's racial mosaic[70]—which promotes the idea, in
effect, that our diversity is enriching and, despite (or because of) our differences,
we are all one people (represented above all by the iconic image of Ovaherero
women with cattle-horn headdresses, a deep irony considering their resistance
to state-sanctioned history)—and heroic images of the war of liberation led by
the South West African People's Organization (SWAPO), mostly represented
by heroic (crossing into the histrionic) battle images.[71] The two narratives came
together in a way such that one, the multicultural mosaic, masked the way that
the other, the heroic liberation struggle, narrowly represented the experience of a
northern political and economic elite. At the same time, the dominant narrative
leaves out of the picture the historical experience and postcolonial justice claims
of those from the southern regions where German colonialism was the dominant
reference point of oppression and the struggle for liberation.

SWAPO has dominated national politics since Namibian independence in
1990, and for most of the postindependence period it has maintained a two-
thirds majority in the National Assembly. Economic inequality ranks among
the world's highest, alongside that of South Africa and Lesotho, with a Bantu
elite joining a small white minority to disproportionately reap the benefits of

extractive industries and agriculture, while programs intended to reduce inequality have consistently failed.[72] Power lies mainly in the hands of the people
belonging to the northern part of the country, Kaokoveld/Kaokoland, Ovambo-
land, Kavango, and Caprivi. These northern territories were largely spared the
destructive consequences of German occupation—the worst effects of colonization occurred in the period of South African domination, with the most intense
oppression, suffering, and warfare concentrated in the struggle for independence
during the late 1970s and 1980s. The period of German rule from 1884 to 1915
did not lead to the loss of land in the northern region. The rebellions against
dispossession and their violent suppression in the form of mass ethnic killing
in the German colony and the thorough postwar expropriation of tribal lands,
in contrast, were concentrated in the "Police Zone" to the south.[73]

Consistent with this historical experience, SWAPO has employed a narrative
emphasizing its central role in the liberation struggle, both to encourage voter
support and to diminish the contributions of political rivals from the south,
mainly Ovaherero and Nama, who fought and suffered in earlier struggles against
colonial rule. The dominant presentation of Namibian history," Zuern notes, was
"a SWAPO-based narrative, in which SWAPO is equated with liberation and support for SWAPO with patriotism."[74] On those occasions when the Ovaherero and
Nama resistance struggles are invoked, they are flagrantly and unapologetically
depicted as struggles for the liberation of the country as a whole.[75]

Hero's Acre (inaugurated in 2002) and the Independence Memorial Museum
(inaugurated in 2014) are the two most prominent expressions of Namibia's state-
sanctioned narrative of national identity, revealing both the dominant narratives
of the state and, over time, the beginnings of their transformation. Both were
constructed by North Korea's Mansudae Overseas Projects, the commercial wing
of an art studio specializing in overseas monuments and memorials. Hero's Acre
is the country's most purely distilled expression of the "hegemonic version of
Namibian history" centered on SWAPO's military struggles against South Africa
during the War of Independence.[76] The memorial, built into an uninhabited hillside some 10 km south of Windhoek, has as its centerpiece an eight-meter-high
"unknown soldier" (bearing a striking resemblance to the first president Sam
Nujoma), holding an assault rifle in one hand and poised in the act of throwing
a grenade with the other. Behind him is a 36-meter obelisk, tapered inward at the
bottom to represent the blade of a sword. The website of the memorial hosted by
the Namibian Tourism Board depicts it as "celebrating the heroes of Namibia . . .
in an effort to never forget the sacrifices and efforts of all the proud Namibians

who fought for freedom and self-determination." This representation of sacrifice and heroism is not intended to inflame grievances but "typifies the Namibian spirit of endeavor and a national psyche of reconciliation with a view to the future instead of dwelling on the past."[77] In an interview with the German newspaper *Die Welt*, Namibian president Sam Nujoma similarly emphasized the need to turn the page on Germany's colonial history in Namibia: "The new Namibia and the new Germany are no longer concerned with the past. We are leaving the sad history behind us and are working progressively together."[78]

As with most memorial practices, there is a distinct presence of absence in Namibia's approach to postcolonial commemoration. The emphasis on "leaving the sad history behind," excluding the history of genocide in the process, was also evident in the diplomacy surrounding a plaque erected on the invitation of the mayor of Bremen, Henning Scherf. The plaque, inaugurated on 21 June 1996 on the occasion of a visit by Nujoma, offers the inscription "To the memory of the victims of German colonial rule in Namibia 1884–1914."[79] There was at that time no objection by the president to the fact that the inscription failed to include the word *genocide* (*Völkermord*); he made no claims during his visit for the return of the skeletal remains of those who had been killed and beheaded in German Southwest Africa and sent off to be stored and studied in German museums and research institutions; and he gave no indication in press coverage of the event that the memorial plaque expressed anything other than an appropriate acknowledgment of the German role in colonial violence.

The Independence Memorial Museum, inaugurated in 2014 (some twelve years after Hero's Acre) maintains an emphasis on the independence struggle led by SWAPO, together with exhibits of photography and artifacts curated with a view toward remembering and condemning the colonial past. It was built on the site of a much-debated German colonial-era equestrian monument, the Reiterdenkmal (which I discuss further below), inaugurated in 1912 to honor the German soldiers and civilians who died in the 1904–1908 war. Like Hero's Acre, the commission for the Independence Museum was given to the North Korean firm Mansudae Overseas Projects and constructed in the style of socialist realism. Mural-sized paintings of the key figures of the independence struggle dominate the exhibits with their monumentality, while less commanding space is dedicated to particular themes or events, such as a bas-relief of a soldier hung from a tree and a replica of Nelson Mandela's cell on Robben Island.

It is noteworthy that outside the museum there is also a memorial acknowledgment of Namibia's history that goes far enough back in time to represent the genocide at the hands of Germany's colonial army. This takes the form of a "geno-cide statue," situated next to Robert Mugabe Avenue on a small hillside outside the museum's main building. In this instance, however, the realist style presents not just a glorification of socialist or nationalist values (a man and a woman in a loose embrace, breaking free from chains) but, featured on the pedestal, the starkest possible imagery of mass ethnic killing. This theme is represented by a bas-relief depicting three people in civilian clothing hanging from a tree while bearded soldiers holding rifles look on dispassionately (figure 8). On the opposite side, facing away from the street, is a bas-relief image of a group of starving Ovaherero, including an elderly woman, with ribs protruding, sunken-eyed, staring toward the viewer. This is the same image from an archive photo that, once digitalized, was widely used by activists, finding a repeated presence—almost a meme—on the web and on T-shirts worn by Ovaherero activists. In both forms of use, the activists seem inured to the horror of the image and are happy to post it or wear it as a symbol of identity. Here it has been set in bronze, making the formal transition from Facebook to monument, from the medium preferred by activists to that preferred by the state.

An inlaid caption above the front-facing image reads, "Their Blood Waters Our Freedom," but, crucially, there is nothing that explicitly connects the monument to the 1904–1908 genocide. When I asked Ovaherero activist Esther Muinjangue about the significance of the genocide memorial, she was dismissive of it. "It means nothing," she said. "It is not even where people can see it. Not like all the other statues."[80] For the Ovaherero activists, Namibia's genocide memorial is little more than state-sponsored lip service in bronze, brick, and stone. In the light of Muinjangue's assessment, the two figures standing atop the pedestal breaking free from chains do seem to offer yet again the government's dominant narrative of looking to the future, not the sad story from the past.

Namibians of German Descent

Under the impetus of the German descendants' sense of their place in the civilization of the wilderness and the founding of the nation, Namibia has experienced its own version of Germany's *Historikerstreit*, a political and intellectual dispute between right-wing scholars who rejected the uniqueness of the Holocaust, stressed the frequency of genocide, and minimized or denied Germany's burden of responsibility for the events of World War II and an opposing side that asserted the opposite, the idea that Nazism was the culmination of Germany's unique political sensibilities and that Germany bore a correspondingly unique

Figure 8 The pedestal of the "genocide statue," Windhoek, Namibia. (Compare figure 6.)

responsibility for its actions. This was a dispute that brought an academic debate into wide prominence, addressing the public through popular media.[81] The Namibian version of this dispute, however, shaded over into full-blown denial of the 1904–1908 war as an event marked by orchestrated, state-sanctioned ethnic mass killing. According to a dominant German-descendant narrative of colonial history, the estimate of 60,000 to 80,000 Ovaherero dead with 16,000 survivors is an outcome of politically motivated falsification; the war involved atrocities committed "on both sides," and the Ovaherero were not at all a helpless, lethargic, or passive people, as evidenced by the atrocities their combatants (and their wives) committed in attacking innocent settlers; and finally, according to this narrative, the notorious "elimination order" pronounced by Lothar von Trotha was not directed to his soldiers, whom he instructed to shoot above the heads of the fleeing rebels, but to the Ovaherero themselves, merely as an act of intimidation that was not meant to be put into practice in the form of mass killing.[82]

The flip side of such revisionism is a celebration of conquest and civilizational development in German Southwest Africa. Consistent with their take on Namibian history, the German descendants have embraced a culture of remembrance that places emphasis on their collective and individual contributions to the greater good of Namibian society. For example, a museum founded in 1951 in the coastal town of Swakopmund—famous for its impeccably preserved colonial-era buildings—features a prominent display of an antique dentist's chair and paraphernalia, a reflection of the profession and passion of the museum's founder, Dr Alfons Weber. Local residents, most with German ancestry, donated the objects that make up the museum's collection with the intention of reflecting their identity and preserving their history.[83] The visitor is first greeted with a display of German brewing equipment, along with explanations of the mastery and good use to which it was put in the hostile land encountered by the settlers. The weaponry and uniforms of German forces in the Police Zone are also prominently and proudly displayed, given much greater prominence than a small, recently installed homage to the 1904–1908 war. With its dentist's chair, as with its other displays, the museum is essentially a forum for colonial nostalgia.

German descendants in Namibia have also been staunch defenders of colonial-era state monuments, erected in honor of those compatriots who died in defense of their colony. Theirs were the loudest voices objecting to the removal of the Reiterdenkmal from the present site of the Independence Museum. One argument drew attention to the fact that the soldier sitting astride his horse offered the first monumental depiction of a commoner—in other words, not a prince or conqueror but an ordinary man doing his patriotic duty. The removal was therefore tantamount to "stealing our history."[84] A similar position against removing the monument, but with a diametrically opposed justification, came from a number of leading Ovaherero politicians from across the political spectrum, who argued that the equestrian monument served as a daily reminder of German colonial-era crimes and should therefore be left in place. It was removed nonetheless, without explicit justification but very likely as an object inconsistent with the dominant historical narrative of the state.

A similar controversy erupted more recently in Swakopmund surrounding the Marine Denkmal, commissioned in 1907 by the Marine Infantry in Kiel, Germany, to commemorate the German First Marine Expedition Corps that fought to suppress the Ovaherero rebellion.[85] It depicts a vigilant German soldier standing on a hilltop, rifle at the ready, with another German soldier fallen below. For years, this monument had been sheltered by its staunch German-descendant supporters, but in early 2016 it received a baptism of red paint, marking its transition to a new kind of monumental personhood: an object of historical indignation. A Google search of the monument reveals the permanence of this act of protest—a solid portion of the images on the web depict it with its red paint, as an object of desecration and loathing. Online, no amount of paint removal and polish can restore it to a condition in which it is not a symbol of oppression and colonial violence.

Ovaherero-Nama Remembrance

The centenary of the beginning of ethnic mass killing in German Southwest Africa in 2004 offered a "hook" for activists and advocates to challenge the Namibian government's SWAPO-centric narrative of the state. They reframed the war as "the 20th century's first genocide" and in doing so took aim not only at Germany's denial of responsibility for its legacy of colonial violence but, perhaps more significantly, challenged the key ideas behind the Namibian ruling party's popular appeal. They drew attention to an entirely different colonial legacy, one that preceded the South African imposition of apartheid and the struggle against it, that involved not only mass killing but wholesale dispossession of those Ovaherero, Nama, Damara, and San peoples who occupied the area of white settlement known in colonial times as the Police Zone.

The boundary separating the Police Zone from the northern areas occupied by the Ovambo and other Bantu-speaking peoples created different administrative zones and dramatically different colonial experiences.[86] The zone was the locus of the most catastrophic forms of colonial domination, dispossession and violence, which the North did not experience to nearly the same extent.[87] Even though the northern-based Ovambo/SWAPO narrative of the state prevailed in public memorials, the memory of the 1904–1908 war has remained very much present in Ovaherero and Nama oral history, reinforced by sites and ceremonies of commemoration.

New media have not entirely displaced the traditions of oral memory. In the Namibian oral histories gathered by Caspar Erichsen and his team of researchers, the emphasis is not always on feats of valor in the battlefield but on heroism in resisting a colonial force bent on destruction. One such story, for example, focuses on Ngatitwe, or Rupertine, as she was known to the Rhenish missionaries among whom she lived, learned, and worked before the war. After the outbreak of hostilities in 1904 she was captured by the Germans and sent to a prison camp in Windhoek. There, a missionary recognized her and, following his pleas to the colonial authorities on her behalf, she was released and returned to her home in Omaruru. Having won her freedom, Rupertine put her missionary education and identity to good use and campaigned against the lynching of captives by the Germans at the Omaruru concentration camp, sometimes even physically intervening by removing nooses from victims' necks. "After Rupertine's intervention," one interviewee recalled, "we never heard of people being hanged again."[88] This was the kind of source material for the oral histories of mass killing and its aftermath that were passed on between generations. The Ovaherero respondents, Erichsen observed, were unlike members of other groups that participated in his oral history project in the extent to which they were "able to track the direct impact of the war on their own families, providing both the names of people who had died and the places where they died."[89]

The Nama too have an oral tradition centered on the events and consequences of mass killing. David Frederick, an eighty-four-year-old descendant of Cornelius Frederick, leader of the Nama rebellion, was recently interviewed by a reporter from the *Wall Street Journal* and used the occasion to encourage German negotiators to visit members of his family and community in Bethanie, a small desert town three hours by car from the former death camp on Shark Island: "They must hear it straight from someone who was affected," he said. "The Nama people, from generation to generation, told the story, and that is the story they need to hear."[90]

The art of memory that Frederick invoked often goes along with material devices that give focus and consistency to recollections. In the case of the Nama community in the Namibian desert, this takes the form of a marble block in the shape of a tombstone, erected by the people of Bethanie to commemorate Cornelius Frederick and 330 others in the community who died with him in the Nama rebellion against the Germans. It was placed opposite a larger memorial to German soldiers, settlers, and nurses who died during the war, many due to illness.[91] Commemorative acts also focus on the grave of Hendrik Witbooi, near the site of the battle in Vaalgras in which he was killed in 1905, with an annual "heroes" day celebration observed in August by Nama-speaking communities.

For the Ovaherero, the central manifestations of communal memory are annual commemorative performances that combine elements of pilgrimage and public spectacle, the *omazemburukiro* (from *okuzemburuka*, "to remember, commemorate").[92] The focal points of these events are the gravesites of chiefs, with different colors in flags and costume representing the various chiefly lineages. White Flag celebrations are held October in honor of the royal family of Wilhelm Zeraua, buried in the Rhenish cemetery in Omaruru; Green Flag celebrations occur in April in Okahandja, the Ovaherero "capital" and in August near Gobabis, a remote community on the edges of the Namibian desert. By far the largest celebration is that organized by the Red Flag Ovaherero, the march to visit the chiefs' graves in Okahandja on the Sunday that most closely precedes the 23 August anniversary of the reburial of Samuel Maherero in 1923. Men in uniform march in formation, some on horseback, in the manner of a military parade. This feature of the event is, as Kössler observes, "an obvious rehearsal of transformed warlike traditions that had been cut off by colonial conquest."[93] Women in lineage colors with cattle horn headdresses and full-length dresses also walk in formation. Aside from its brochure-ready tourist-attracting spectacular qualities, the annual event is an occasion for public speeches that mark the traumatic experiences of the 1904–1908 genocide as a common source of collective identity and, perhaps more significantly, that loudly and publicly proclaim demands for recognition and reparations. It has also made a transition to social media, becoming a reference point for concentrations of pride and posts. It has done this in much the same way that graffiti has transcended the genre of wall writing to become a dual medium, with offline and online forms of creativity, with their own audiences.

The resurgence of Ovaherero identity as public spectacle is not restricted to the parades and ceremonies that take place annually in late August; it has also taken on a sharper edge in protests focused on colonial monuments. In July 2008, fifty-one wooden crosses were set up around the Reiterdenkmal, bearing the names of victims and expressions of condolence and remembrance in Otjiherero. In October that same year a Namibian flag was placed in the rider's rifle barrel. Every time that protesters materially resignified the statue, discussions would flare up in local news media and on social media about the status and justification of a monument erected to glorify German colonialism and its violence.[94]

The Namibian government appears to be acutely aware of the political challenge inherent in the Ovaherero's and Nama's common resurgent identity as the victim-descendants of the twentieth century's first genocide. In political-ideological terms, this identity poses a direct challenge to the dominant SWAPO-oriented narrative of Namibia's independence. In response, the government has engaged in the combined politics of strategic avoidance and of co-optation.[95] Strategic avoidance can be seen in the fact that government officials pointedly avoided participating in the events commemorating the 2004 centenary of beginning of the colonial war. Co-optation can be seen in the activities surrounding the government-sanctioned "Heroes Day," which commemorates the beginning of Namibia's war of independence, in which the first key event took place on 26 August 1966 at Omugulugwombashe. It is marked in the usual way by speeches, wreath laying, and the bestowing of national honors, such as military medals, but its emphasis on "heroes" as a name and a focal point is significant, one of many connections and points of overlap between the national holiday and the Ovaherero's celebration of their reconstituted identity. The effect of this overlap is to fold the biggest commemorative event of a political rival into a national holiday focused on celebrating the foundations of the state.

The Ancestors Speak

Co-optation has also begun to extend to the government's handling of the Ovaherero's and Nama's genocide claims. On 30 September 2011 the medical research facility Berlin Charité repatriated twenty human skulls, eleven identified as Nama and nine as Ovaherero, to an official delegation from the Republic of Namibia. Crowds thronged the airport in Windhoek to welcome the delegation and their precious cargo. The skulls were a small fraction of thousands (to this day, no one knows the exact figure) of remains of Ovaherero and Nama victims of colonial violence in German Southwest Africa that were stolen and sent to Germany for purposes of research, much of it oriented toward proving racial theories of European superiority. The German government created a scandal, however, when Minister of State Cornelia Pieper failed to acknowledge the genocide or apologize for it in the name of the German nation and state and left the room before the end of a meeting with the Namibian delegates.[96] Activists in both Germany and Namibia with whom I spoke continue to point to this "missed opportunity" (as one of them put it) as a low point in Germany's response to their justice claims.

The eleven skulls identified as being of Nama origin that were returned from the Berlin Charité medical complex were not given a place of burial near Shark Island, as the descendants of Cornelius Frederick had expected, but were placed in storage at the Independence Museum—returned, in other words, to a site of state power as symbols of colonial history rather than as "ancestors" to those who trace an intergenerational connection with them.[97] With over a thousand museums in Germany and no legislation coordinating their obligations toward the descendants of colonial occupation, there is no clear understanding of how many remains are scattered throughout the country or where they are located. Skulls with provenance in the colonial genocide have even recently been discovered in Strasbourg, a former German city presently in France. A small number of skulls was repatriated from Germany to Namibia in secrecy in 2014, without the knowledge or participation of the Ovaherero or Nama leadership. A further repatriation of skulls, now stored in the Staatliche Kunstsammlungen Dresden (SKD), is soon to take place. Where and how they will be received is anyone's guess.

Other remains have been found overseas. In 2018, Ovaherero skulls were found in the American Museum of Natural History, where they had been acquired from an Austrian antiquities dealer, Felix von Luschan, at the time of the active shipment and trade of bones originating from German Southwest Africa. This discovery prompted a debate among the Ovaherero themselves as to whether they should be immediately repatriated or kept in New York as a way to make the context of genocide in which some of the skulls were taken more widely known.[98] For the time being at least, the skulls will remain where they are.

The online postcolonial campaign in Germany became a transnational conduit for mobile value systems. These mainly took the form of reappraising colonial history through the lens of justice claims, recognizing the 1904–1908 genocide, and, despite the passage of time, dealing with the Ovaherero and Nama as legitimate victim groups. There was, however, another kind of travelling model that followed from the campaign's main source of leverage. The moral claims the Ovaherero and Nama made against German museums and research institutions for housing the remains of the ancestors killed in a genocide also provided a pathway for structures of chiefly power and veneration of the dead to transplant into settings normally associated with order and rationalism. Pressured by a grassroots social media campaign for repatriation, some of the major museums and research institutes in Germany were compelled to be receptive to the feelings of claimants and to see the bones they kept in their facilities as having personhood. The institutions that once viewed the remains of the Ovaherero and Nama as "research material," were compelled, through the persuasive force of the campaign to recognize the genocide in German Southwest Africa, to see them as human and inseparable from living descendant-claimants. Paradoxically, it was owing to their digital presence that the bones could be recognized and given life.

Museums that repatriated the remains of genocide victims were also witness to ritual acts of solemnity and respect that connected the ancestors to their living descendants. Among the Ovaherero in Namibia, funerary processions are a foundation of collective belonging, going back to the 1923 return of the remains of Samuel Maherero, from Bechuanaland to Okahandja in the heart of the then-fallen German empire. Following from this foundational event, funerary processions acquired the added significance of postgenocide collective reconstitution. Human remains are the conduit for an active connection between the ancestors and their living descendants, mediated by chiefly ritual authority. The Ovaherero adopted the rituals and attitudes associated with the ancestors in the new settings in which remains had been found and were being prepared for return. The repatriation ceremonies of 2011 included Ovaherero chiefs who performed the rituals of their lineages, kneeling, heads bowed on the steps of the church where the skulls were assembled waiting for them.

More important, the skulls and bones have reciprocal agency. When she learned of the existence of Ovaherero skulls in the American Museum of Natural History in New York, Esther Utjiua Muinjangue was thrilled. "Now we have the ancestors there to guide us in our court case," she told me, her eyes sparkling. More solemnly, Nama activist Talita Uinuses went off script from her speaking notes in a symposium on repatriation at the Grassi Museum in Leipzig and said, "There are skulls of my ancestors in this building. I feel the vibrations. All day long, I have been feeling the vibrations."[99] In spontaneous moments like these, a difficult-to-express quality of the conflict over the remains of the ancestors makes itself manifest: an active line of communication between the departed and the living through the relics of the dead.

Truth and Power

> There never is any such thing as one truth to be found in dramatic art. There are many. These truths challenge each other, recoil from each other, reflect each other, ignore each other, tease each other, are blind to each other. . . . Political language, as used by politicians, does not venture into any of this territory since the majority of politicians, on the evidence available to us, are interested not in truth but in power and in the maintenance of that power. To maintain that power it is essential that people remain in ignorance, that they live in ignorance of the truth, even the truth of their own lives.
>
> Harold Pinter, Nobel acceptance speech, 2005

The Problem of Knowledge

Human rights have always relied, explicitly or implicitly, on the premise that factually exposing the conditions of mass harm will one way or another provide the means toward its remedy. Legal instruments, structures, and processes of international law do of course perform much of this task, with meetings of the Universal Periodic Review (a mechanism of the UN Human Rights Council) constituting the core of the UN's oversight and compliance regime. But there are no incentives toward compliance with human rights that do not ultimately turn to public exposure of the crimes committed by states, the familiar "sunlight" effect, "naming and shaming," and the "politics of embarrassment." Compliance is contingent on mass persuasion and the political costs that follow from the exposure of acts that violate the ethical standards that states have agreed to uphold. For this reason, human rights have always been inextricably connected to technologies of information and communication. The changing means by which information is recorded and communicated on a mass scale constitute the pathways of exposure, public persuasion, sympathy, indignation, and reckoning that are central to human rights compliance.

In this book, I have approached this close connection between human rights and information technologies (ITs) as occurring in three historical phases. In what I call the analog period of human rights, journalism was *the* original strategic reference point of activist efforts. The first mass exposure of the crimes of states in the postwar era occurred through the intercession of journalists working for broadsheet newspapers or major television networks. In the major channels of communication with publics, there was little place for activists, and those who did engage in justice campaigns did so with an eye to persuading journalists to report on their cause. The events they organized and the information they assembled were prepared in such a way as to attract the attention of reporters, produce mass coverage, and shift public discourse. This was easier said than done. Journalists of the kind that had wide appeal were not inclined to take up just any story; activists had to be creative to get their attention. The work of Serge and Beate Klarsfeld, including a street protest they organized in which Holocaust survivors participated in unfurling a Nazi flag under the apartment window of the war criminal Kurt Lischka, provides an early example of such impossible-to-ignore, press-garnering creativity in the assertion of human rights claims.

With the development of mass-consumer digital technologies in the era of "netizen" activism of the 1990s and early 2000s, it became possible to bypass journalistic filters in campaigns of public outreach that navigated connections within and between online communities. Arguably the most important change brought about by new ITs in the first digitally networked era of human rights was the end of the monopoly of journalistic filters in mass communication to publics. Individuals and organizations with collective grievances were potentially able to bypass the gatekeepers of information and directly reach mass publics through new media. With the development and wide accessibility of the web, the process of representing a claim, cause, or community of rights holders was not limited to journalists of mainstream media outlets or high-profile NGOs with human rights agendas, but went to the source: to communities of claimants and their immediate evidence and discourse of harm. Some found hope in the rise of the internet's networked society in much the same way that hope was first attached to human rights aspirations in the postwar era—as a rational way to build a better world. There would be no stopping the momentum of truth exposure untethered from structures of propaganda and state-controlled media.

Now, however, things are not so sure. The new era that I refer to as human rights 3.0 is defined more than anything else by transformations taking place in truth claims, above all the rise of rights-oppositional processes of post-truth obfuscation, state-sanctioned institutions of censorship, and powerful and largely unaccountable systems of data acquisition and processing. Transformations in the pathways of justice claims are ultimately situated in new big data and cloud computing technologies of information and communication. With the growing power and reach of these technologies, the forces of repression have shifted more than ever before toward control of knowledge, inscrutability, ignorance, and deceit in the service of the powerful.

In a sense, the uncertainty over knowledge is moving people toward struggles with meaning-making that are more familiar from much earlier times. Like the heretics of the late Reformation—those who elaborated millenarian conceptions of a new heaven and a new earth in the context of popular translations of the Bible and a technological revolution in typography—there are those today who are suspicious of received orthodoxies and in their stead are inclined to blindly trust conspiracy theories and doubts within small communities of the faithful. And like the dissent and insecurities of earlier times, there is a ferment of inquiry, a search for the forms and foundations of knowledge that can be trusted. This is an increasingly philosophical era, in which the age-old question "How do I know what I know?" is being asked with greater frequency by greater numbers of people than ever before.

Human rights, with their reliance on evidence of mass crimes perpetrated by states, are an ideal window into this revolution. They bring focused attention to abuses of knowledge, to the confusion produced by ideologically inspired unmooring from facts, and to the truth-exposing capacities (and limitations) of new technologies.

In this newest era of human rights, civil society is being thrown into disarray. State-sanctioned programs of censorship and disinformation make use of new ITs to control information and opinion, in violation of those rights intended to protect the foundations of political freedom. At the same time, there is an almost self-destructive quality to the decline of effective public discourse and dissent. Social media's reinforcement of online communities is built on a narrow bandwidth of affirmation and conviction that encourages uncompromising and unsympathetic opinion—a phenomenon that has profound implications for the justice claims and causes of those who are displaced, marginalized, and subjected to violence. In this media environment, emotions readily overtake facts as a central criterion of truth. History becomes therapeutic, a source of affirmation rather than collective self-examination. The spread of disinformation on a mass scale has become an obstacle to the assertion of rights claims, in which establishing the truth of conditions of violence and loss of political freedom is a precondition for attention and intervention.

Noting this decline of civility, some have looked to intrinsic qualities of social media as platforms that magnify the enclosures and vituperations that in other circumstances would have had little reach or influence.[1] The concept of the "filter bubble" or "echo chamber" considers the effects of group enclosure and ideological positioning facilitated by online platforms.[2] The rise of post-truth as a political strategy and technique for muddying the water of justice claims has upended the capacity of marginalized groups to use victim testimony as a pathway to recognition and remedy. The media being used as vehicles for communicating injustice and soliciting the responses of indignation and sympathy are in themselves conducive to the narrative oversimplifications and hatreds at the root of mass violence.

At the same time, largely because of the unreliability of public discourse, new media have added to the appeal of forensic evidence as a foundation for public sympathy. Digitally anchored pathways to knowledge are becoming more central to justice claims and the provocation of outrage in mass publics—including that more exclusive and powerful public in diplomatic and global governance circles known as "the community of nations." For example, it was not so much the cultural color and life of the markets and mosques of Aleppo that made the city's destruction by the Syrian government an offense to world opinion. Rather, it was Syria's use of internationally prohibited cluster bombs and chemical weapons—the remnants of which (in part through the efforts of the Bellingcat investigative team discussed in chapter 3) could be incontrovertibly proven from images of the aftermath of air strikes, uploaded onto the web. The suffering of women and children was of course a source of sympathy and indignation, but it could be attributed to any one of the warring parties. A serial number on a piece of metal in the rubble of a building, however, could not be quite so easily dismissed. In the reduction of claims to forensically assembled facts, we are witness to a kind of "bare death," with mass atrocity moving people to sympathy through little more than image-assemblages of shrapnel and bomb parts.

Taken together, the material I have presented in this book forms the contours of a dilemma: the only effective way to counter state-sponsored disinformation campaigns and the pervasiveness of post-truth conspiracy theories is a turn toward quantification and "factual" material evidence. This includes the use of indicators and new technologies of data processing to offer consumers of injustice a sense of informational security. But quantified representation of information is inherently abstract and distant from the images and forms of narrative that people find appealing. Numbers alone cannot tell stories. Algorithms are an abstraction beyond human reckoning. The sense of injustice is not activated by facts of this kind. It prefers emotional involvement, glimpses of reality, the horrors of violence, identifiable individuals (preferably innocents) and their lives as victims, and the poignant romance of disappearing ways of life.

Technologists as Activists

Stepping in to meet the need for knowledge security in conditions of post-truth are the experts who specialize in collecting, processing, and analyzing data. In human rights 3.0, new technologies are moving activism toward a greater reliance on specialized technical knowledge, institutional infrastructural support, and donor financing. These conditions are producing new forms of advocacy and identity that are increasingly shaped by educationally and technologically empowered elites.

Technology corporations are central to this shift toward specialized expertise. They alone possess many of the proprietary tools for processing the large, complex data sets known as big data. Their participation in global governance initiatives in collaboration with UN agencies brings with it the information derived from social media, and the human, technological, and infrastructural resources to gather and process this data on a scale never before seen. Their commitment to global governance as a portion of their research programs overall may be miniscule, but they bring resources of a kind and scale that are (for the moment) not available elsewhere. This is taking place in the context of a growing movement toward corporate responsibility for human rights, not only in the tech sector but also in other areas in which industrial activity has major consequences for freedom, security, and environmental sustainability.

There is, however, something schizophrenic about corporate participation in human rights and in the UN's Forum on Business and Human Rights in particular. The annual gathering of some two thousand participants hosted by the UN Office of the High Commissioner for Human Rights is premised on a fundamental moral tension. On the one hand, it is drawing more public attention to abysmal corporate human rights records—criminality and atrocity in the absence of the weapons of war—even though the international legal structure of these rights is oriented toward states. And on the other hand, they are concerned with corporate image as it might affect profitability and wish to construct an image or "culture" of rights compliance and justice advocacy. The tech sector is thus situating itself as a defender of human rights by putting its resources behind new big data projects in collaboration with the UN. The transmitting and recording technologies of drones and satellites are serving to map out the sites of conflict and provide evidence of war crimes. Satellite data, combined with AI, constitute powerful forensic tools able to track the movements of military operations, record evidence of atrocities as they happen, and identify likely locations of mass graves in the aftermath. Big tech corporations are in a position to act both like an evil entity of science fiction imaginings and like Boy Scouts.

There is something substantively different about private sector engagement in human rights initiatives, which would have been unrecognizable in the analog era. The paradoxical effects of corporations' participation in human rights initiatives can be seen in the simultaneous conditions they create for furthering the UN's goals of, say, service delivery to refugees, and yet they are responsible for the systemic corruption of the goals of transparency. In the use of large-scale data sets, the collection of metadata (or data about data) and the use of algorithms and machine learning (possibly with computers writing their own code) have made the sources of decision-making largely unknowable and unexplainable, beyond examination and possible remedy. Tech experts are universally aware that the usefulness and accuracy of a software program depends on the quality of the information and assumptions at the entry point of its code. Yet many users, including judges and policy experts in global governance, rely on them in ways that are incommensurate with their potentially flawed origins. The result is growing use of bureaucratic tools premised on infallibility. The technological instruments being applied toward human rights remedy have built-in proclivities toward human rights violation.

One outcome of the prestige and rewards directed to software engineering skills, with their connection to new forms of both NGO activism and corporate profitability and power, is a fetishization of software development in human rights advocacy, to the detriment of other roles. In human rights 3.0, dissent is increasingly driven by those who have insight into the digital structures and sources (or source codes) of oppression derived from specialized training. Simply put, we now rely irreversibly on technical experts for the defense of freedom.

This means that possibilities for the "vernacularization" of human rights principles and local use of NGOs as a form of organization (however problematic these might have been in the era of networked activism) are being displaced by expert-driven technologies and ideals for digital intervention that cannot be readily translated or simplified for widespread use. There may be some truth to one tech expert's casual observation to me that "anyone can write an .exe [executable file]" using off-the-shelf software that puts users beyond the biases and prejudices of controlled media. Certainly this skill has become more accessible than ever before. But this perspective does not reflect the global reality of the way that expertise is distributed. If it is true that roughly 2.5 billion people do not have a bank account, it can be reasonably assumed that the expertise required to write an .exe is still fairly unusual. What is more, making use of ready-made programming software is very different from the level of expertise required to work at the forefront of the digital arms race against state uses of new technologies to encroach on privacy and civic freedoms. In human rights 3.0, the specialized knowledge needed to be an influential actor is simply beyond the capacities of the vast majority of those who are the central subjects of human rights.

This has important implications for the global project of human rights. Not long ago the main difficulty for the universalization of human right norms used to be situated in the chain of transmission through intermediaries. Now it is the fact that in key realms of rights violation, technical specialization has broken the chain of transmission altogether. For the time being at least, there can be no mass communication of universal norms when the instrumental expression of those norms is in code.

Participatory Technologies

We must be careful, however, not to assume a stark boundary between educated elites and ordinary justice claimants. The Stanford Solidarity Network—the student activist movement situated in tech education—is making use of the same kinds of petitions and student organization techniques widely used during the civil rights movement and the Vietnam War protests of the 1960s and 1970s, except that today the focal point of their solidarity is an online petition (MoveOn. org), in common with many others seeking support for important causes. Their expressions of dissent might gain leverage from their status as the future of software engineering, a rare human resource in the more specialized domains of technological development, but their strategies of protest and outreach are accessible and familiar. If their protest catches on, they would have a great deal to say about big tech's record of rights compliance and conditions of transparency to their employees.

Considering human rights claims and forensics processes more broadly, it becomes clear that human rights advocacy is not entirely in the hands of a technologically empowered elite. Following the fact that computation occurs through both assemblages of data and structures of processing, the evidentiary foundations of human rights claims correspondingly fall into two very different categories: first, the masses of data accumulated through digital mining (such as in drawing data from social media platforms) and second, the computational programs used to process this data to reach conclusions and assist in decision-making. Corresponding with this division in the structure of computation, the evidentiary foundations of human rights are being revolutionized by new technologies at opposite ends of a spectrum of specialization: by the elite, highly educated designers and programmers of software and by the ordinary users who serve in processes of "citizen witnessing" and participatory fact-finding.

Besides the sophisticated, specialized uses of new technologies, there is an important participatory dimension to the accumulations of digital data. As portable devices that combine the technologies of photography, video, audio recording, and typography with what appear to be a never-ending supply of applications, smartphones are fomenting a revolution in the way humans see, interact with, and "witness" the world. The near ubiquity of these devices means that acts of mass violence are often recorded one way or another, either in the act or in their immediate aftermath. Their uses include strategies involving the creation of what Myron Dewey, founder of the web platform Digital Smoke Signals, calls "visual legal narratives," the evidence produced by the simultaneous recording and broadcast of events as they happen. Livestreamed video evidence is seen by some activists as a pathway to secure knowledge. There can be little doubt about the truth of an event when it is unfolding before one's eyes, together with thousands of others viewing online. In being streamed, conflicts with agents of states and corporations, as in the Dakota Pipeline standoff in the Pine Ridge Reservation, are subjected to immediate public exposure, and with it, circumvention of post-truth denials and obfuscations. The empowerment that came with the rise of internet activism and networking in the 1990s has not been entirely undone by censorship and misinformation. State-sanctioned efforts to control the internet are, as in China, effective to a degree. But with the near ubiquity of smartphones, there is now a greater presence of citizen witnessing. "Victims" have become visible without need of intermediaries and are increasingly able to take on the role of "activists."

Some NGOs, armed with sophisticated technical knowledge and tools, are making active use of these participatory resources. Web scraping—gathering and copying data from the web for later analysis—is another new source of forensic evidence and knowledge security. Bellingcat, a collective dedicated to upholding human rights (and law more generally) through use of publicly accessible, open source technologies of data mining and processing, (an example I discuss at greater length in chapter 3) has directly taken up the challenge of using online data to expose crimes for which states are responsible. Bellingcat has an interesting two-pronged model of intervention: one makes use of technical expertise and innovation and the other communicates this technical knowledge to activists in regular workshops in Europe and North America. Techniques worthy of state intelligence agencies are being passed on to a broad base of information-savvy activists.

Given the educational empowerment that is the beating heart of technological innovation, universities would be expected to take the lead in developing technological innovation in furtherance of this dimension human rights. In fact, some promising initiatives along these lines are currently under way. Cambridge University's The Whistle, for example, is a platform under development (at the alpha stage as I write) that would provide a way to verify and make secure the data from "citizen witnesses" of human rights violations. And the Berkman Klein Center at Harvard is developing an Internet Monitor initiative (currently in closed beta and several years from release) that would let users test websites to confirm whether they are being blocked. These are two instances among a small handful of academic start-ups oriented to the protection of human rights defenders, and it remains to be seen if more university-based resources will be dedicated to technological empowerment of this kind.

Participatory technologies have transformed human rights monitoring beyond recognition. Not so very long ago, the information pathways of human rights went from "headquarters" to the "field" in the form of highly trained specialists, well adapted to travel but not necessarily to the languages or social circumstances of their assignments. Technology has reversed the flow of information so that it now moves more in both directions. "Upstreaming"—in which knowledge derives from "below" in ways that ultimately influence the structures and concepts of UN agencies[3]—now involves the victims and subjects of human rights violations who are finding a place in the otherwise technical knowledge structures of monitoring and forensics.

Claims Making and Curation of the Self

For all the attention given to his accounts of media revolutions, Marshall McLuhan more than anyone else is responsible for the idea that a new medium does not always replace older ones but enhances their subtler properties in ways that bring out some of their overlooked capacities. "New media," he said, "do not replace each other, they complicate each other."[4] He referred to this paradox as the "rear-view mirror effect," meaning that old media continue to serve as points of reference for innovation by being re-elaborated and integrated into new technologies in new combinations and hybrids.[5]

This phenomenon can be seen in the material I presented in chapter 1, in the rise of new kinds of graffiti and the ideas expressed in them. The study of graffiti lends itself to the simple observation that in the pursuit of justice, media rarely stand alone. One of the oldest forms of popular expression has adapted to the internet in ways that are visible in the justice claims sprayed and affixed on walls, sidewalks, bridges, lampposts, and anywhere else that public space lends itself to a message being seen by passersby. The URLs that are common on sticker graffiti are a way to pursue cursory messages into the web for elaboration, with the stickers serving as "teasers," or provocations that inspire curiosity. There is a reciprocal relationship between the public graffiti message that can be acci- dentally encountered and the information online that offers less attachment to place and more informational detail. Old technologies do not always disappear but are complicated and repurposed by the new.

Graffiti oriented to justice claims can be found, by those with an active eye for such things, in many, often unexpected places in the urban environment, but they tend to concentrate in locations that reinforce messages of rights violation, offenses to dignity, and denial of justice. The area outside the main gate of the European Court of Human Rights in Strasbourg offers an example of such a site of concentrated outreach, both to the public made up of pass- ersby and to the specific audience of officials, lawyers, and judges of the court. Public space is a locus of claims making through technologies both in their simplest form and using digital media, linked to online sites of information in the service of a cause.

There is a tendency to abstract out and lend attention to appealing new technologies with far-reaching impacts and to overlook their contextualization in many other activities and strategies of dissent, some involving nothing more than street presence, placards, and "voices raised." Transmedia organizing, or transmedia campaigning, reflects this diversity of strategies.[6] It is oriented toward shaping opinion through a cumulative mass of reflections, encounters, and cul- tivated emotions, communicated through any available media, with influencing public opinion as a central common goal.

Considering human rights advocacy from the ground up, starting with the origin of conflicts and causes behind human rights violation and moving toward activist strategies and struggles toward remedy, provides some nuance to the view that new ITs offer revolutionary tools for social justice advocacy. In chapters 5 and 6 I present two in-depth examples of human rights campaigns that drew on new possibilities for mass communication and self-representation: the Tuareg pursuit of regional autonomy in the Central Sahara and the Ovaherero-Nama claims toward recognition of and reparations for historical genocide. Each of these campaigns struggled in different ways and for different reasons to overcome public apathy and resistance, and in the process, each draws our attention to qualities of the new media environment that, only to a degree, overcome invisibility and marginalization, leaving the activists somewhere at an insecure midpoint on the arc of justice.

There was a major difficulty facing those Tuaregs who were trying to make a rights-based argument for an honorably negotiated peace with regional autonomy in the central Sahara in a way that would convince the distant consumers of violence imagery: those who were lobbying for human rights and negotiated peace had the least political authority in the desert, while those who were better than anyone else at desert warfare were the least willing to make peace-oriented compromise and, even if they did advocate for peace, were the least capable of shaping public opinion toward it. How does one make a public case for self-determination as a necessary solution to violent state centralization when the people one represents have their own deep and complex history of violence as a means to exercise political will? This observation applies particularly to those who pursue their rights as an indigenous people because publics tend to have heightened expectations that they will conform to ideals of timeless wisdom, peacefulness, and cultural color.

The war in northern Mali in 2012–2013 took place in a media landscape as perilous as the desert in which the fighting took place. With mainstream media coverage focused on the simple narratives of jihadism and violence, the Tuareg rights activists were left on their own to shape the telling of their story, with their own uses of media technology. New ITs were put to use in the service of claims to self-determination. In the age of Facebook, Instagram, Twitter, YouTube, and WhatsApp, Tuareg youth could make their attachments to Azawad visible to people in their extensive networks. But justice causes disappear outside these networks if those who fall victim to violence do so in contexts of political and ideological complexity. Matters were certainly not helped by false-flag conspiracies and the shifting alliances of armed groups. As an outcome of the new constellation of violent factions in Tuareg society, the images of desert war-fare—the heavily armed men with turbans, AKs, and Toyota Hiluxes—came to dominate both the mainstream media and popular understandings of the central Saharan conflict. These images overwhelmed the purely abstract descriptions of the state's unmet rights obligations and hopes for a peaceful homeland.

In the Ovaherero and Nama claims of colonial genocide, efforts to provoke public sympathy ran up against a different obstacle: the limits of memory and the public demand for personification of victimhood. With no living survivors of the 1904–1908 genocide, there was no one to represent the violence, to appeal directly to the public with their stories and emotions. Intergenerational claims of genocide confronted the difficulty of retroactive application of law (the Genocide Convention was adopted in 1948, some forty years after the end of the genocidal violence in German Southwest Africa); the claims also hit the wall marking the limits of historical contrition. The German public had great difficulty extending its habitus of recognition and atonement for mass violence beyond World War II and into the sordid history of colonial atrocities.

Where the Ovaherero and Nama did reach people was through exposure of the scandal of state-funded museums and research institutions holding the human remains collected in the context of the genocide, from among those killed in mass executions and death camps. Making this claim was a two-step process. First, the facts of the genocide had to be established, even in the absence of a political will toward reparation. With the terminology of *Völkermord* (literally, murder of peoples) spoken from the mouths of politicians, the bones shut away in the drawers and lockers of German museums and research institutions took on a different character. They were no longer scientific specimens contributing to the project of human knowledge and progress, but gruesome trophies of mass killing, ancestors of living people bereft of their dead.

This was a claim that lent itself to social media. Facebook became the platform of choice for public outreach through regular postings of archival images of ema-ciated genocide survivors in chains and images of the skulls (the quintessential captivating symbol of death) alongside the museums where they were kept. This added poignancy and immediacy to the Ovaherero's and Nama's grievances, bringing them into the present. Facebook defied the monopoly tendencies of museums as custodians of the history of the state. The social medium with the transgressive capacity to construct identities and navigate networks through the communication of grievances confronted the states' monumentalist displays of power and pride of heritage. The story of the nation built into Germany's museums and monuments was contested by the counternarrative of its victims, as told through digital platforms based on quasi-utopian ideals of a connected humanity.

This case illustrates some of the ways that digital media facilitate transnational flows of images, ideas, and indignation in which diasporic activists actively draw from relationships, rituals, and information from home, even (or especially) when the government of their country of origin is hostile to their efforts. This "travelling model" dynamic of justice lobbying can produce not only immediacy to campaigns on the other side of the world but also prod and influence behind-the-scenes points of diplomatic contest and claims making among states, as manifested, for example, in ongoing closed-door Namibia-Germany discussions of reparations for the 1904–1908 genocide.

The transnational dimension of the Ovaherero-Nama campaign makes clear that enabling conditions have arisen for grievances to travel, for claims to translate into new settings and to draw inspiration from new collaborations and networks. In this sense, the mobilizations of grievances have much wider significance than the immediate controversies at their origins. In some circumstances, they reveal the reinforcement of identity attachments and political leverage achieved by postcolonial justice campaigns. Without this transnational dimension to their claims, the Ovaherero and Nama would likely have been subject to the dominant and exclusionary historical narrative of a Namibian government with a secure hold on power. But by drawing from the energy, expertise, and resources of a postcolonial movement in Germany that was striving to push the limits of "memory culture" further into the past, Namibian activists gained additional influence in their efforts to reframe the legacy of German imperialism in Africa. At the same time, the coalition of postcolonial activists in Germany, through their connections with the Ovaherero and Nama, could introduce a compelling case of genocide, plunder, and research ethics gone awry in their efforts to gain wider attention to the contested spaces of colonial memory. By drawing public attention to memorials, human remains, and other material invocations of colonial history, the Ovaherero and Nama campaign for justice gained the kind of traction that it needed in the absence of witness narratives. It did this through new forms of transnational collaboration, in which the challenge of asserting an alternative narrative of the state in Namibia was complemented by postcolonial activism and its goal of adding depth to the narrative of the crimes of the state in Germany.

In each of these case studies, the difficulties of conveying moral complexity through representation of justice causes and their claimants were a central challenge to human rights advocacy. In the case of the Tuaregs, this difficulty followed from the complexities of their history, leadership structure, and radically divergent values; for the Ovaherero and Nama it was because the genocide that is the foundation of their claims is deep in the past, beyond the memories of living witnesses.

Binaries and Enclosures

Why are we able to tell complicated stories in art but have more trouble seeing (and assigning our sympathies to) the complexities of actual conflicts? Why do victim-perpetrator binaries readily get broken down in film or novels, where they are appreciated, while these binaries come to the fore and wreak havoc with efforts to overcome violence or come to terms with historical atrocities? In all these cases, the viewer/reader sympathizes *because* the characters are morally complex in ways that Harold Pinter points to in the epigraph of this chapter, in ways that we have inherited from the storytelling of the ancient world. In light of this, perhaps the right framing of the question should be, Why do publics so commonly succumb to (or demand) stereotypes in the stories they consume relative to human rights violations? What is it about appeals to justice that makes them susceptible to oversimplification of the characters involved?

In approaching this question, one of our considerations should be the market model of human rights campaigning, which frames justice causes as products to be pitched. The arts of cinema and literature are of course also consumer products, but of a different kind; the product is entertainment or edification but with the story as an end in itself. In justice campaigns the goal sought from the consumer is different; it is the sympathetic response and the call to action, the kind of cultivated empathy that is the pathway to the "add your name here" and "donate now" buttons.

The process of competing against others for public attention, sympathy, and activist support in the context of the silencing effect of noise has implications that go beyond the particular causes and claims at the center of legal contests. It has dramatically transformed the way that human life is commonly understood. The internet has introduced a new power applied toward the construction of memory and selfhood, inviting its users to reflect on their essence and to curate it before presenting it to a global audience. The internet's displays of collective being can be seen as free-flowing manifestations of a new stratagem of release from political repression, one that emphasizes the public expression of strivings toward self-determination, sharpened boundaries of community identity, and intensified pursuits of autonomy. Because of its multimedia potential, it encourages the display of historical and cultural artifacts, digital exhibitions in which the most impressive, representative relics of collective being are selected for display, while jumbles of imperfect, ignoble artifacts are locked away from public view. Ideas that appeal across the widest spectrum of attention and sympathy become those most likely to be posted as expressing the essence of a group's, community's, or people's innermost being. It is therefore a means toward ideas of collective injustice and human rights claims. At the same time, it has an affinity with nationalism and strivings toward self-determination—sometimes putting human rights and nationalism in uncomfortable proximity.

It is when we combine the effects of technologies of communication with technologies of law that the new era of justice claims and campaigns comes into sharper focus. Law has its own proclivities toward enclosure. It simply will not do to for a lawyer representing collective rights claimants to say that an injustice affects anyone who subjectively thinks they might have been harmed by a wrongful act. The list of claimants in a lawsuit has to be closed; you are either a plaintiff who stands to benefit from a judgment or you are not. Law seeks (though it rarely finds) certainty and finality and will even distort reality to achieve it.

An analogous form of enclosure applies to collective claimants of human rights. When a minority at risk (MAR) makes a plea for protection from deportation and mass violence, a number of questions follow: What is it about this minority that makes it a minority? What are the differences that serve as the foundation of a claim of collective difference? Who belongs to it and who doesn't? Add the element of public recognition to these questions and the central issue becomes, What do these people practice, believe, or say that makes them both different and compelling? What makes them *worthy* of protection? (This question often comes with the subtext "In your response, please entertain us.") There is a technology (or *techne* in the sense of a "craft" or "art") of public outreach and identity, in which the attributes that make a people worthy of protection are publicly negotiated in a back-and-forth between collective claims and attributes and public response: approval, sympathy, and participatory action or disapproval and silence.

The absence of a response is sometimes just as telling as the plea or provocation that attracts public sympathy. One of the consequences of the polarization of political identities in North America and Europe—with the radical fringes on the right and the left amplifying the language of division and intolerance—is that persuadable and possibly sympathetic publics are increasingly reluctant to throw themselves behind anything that might be seen as identitarian politics or collective rights claims. In circumstances in which the truth-value of facts has become secondary to the value of feelings—and in which the majority views this development with dismay—there is much less to be gained from asserting a claim that one's *identity* has been violated and stands in need of protection.

Finally, let me revisit the question of technological determinism in light of the attention I have given to human rights. To what extent are human rights shaped by the media through which claims are communicated? An argument can be made that the growth of human rights instruments that the world witnessed in the first decades of the twenty-first century was an outcome of the identities and indignations that coalesced in the 1990s, with the delay attributable to the time it takes for new human rights instruments to be negotiated and ratified. And this coalescing of identities around such reference points as disability, gender identity, and indigeneity, in turn, can be attributed to the human connections and forms of self-representation facilitated by the Web 1.0. There is in this sense a reasonably direct connection between new ITs and the design of the normative frameworks that constitute human rights in international public law.

Things get a bit more complicated when we shift our focus to the publics that are the targets of claimant outreach and the major source of leverage in processes of human rights compliance. Has the hegemony of human rights persuasion changed at all under the influence of new, farther-reaching information technologies? And if so, in what ways? Here the answer is unclear, or to the extent that it is clear, it takes the form of a contest with an uncertain outcome, between those advocating for human rights and those seeking to control opinion and dissent. Powerful new tools—smartphones—have put witnessing and forensics in the hands of billions; meanwhile, repressive states are investing considerable resources to shape opinion, manage those tools, and cast doubt on those who use them.

This is a contest of technologies, yes, but it is also a contest of the underlying values to which new tools of information and communication are being applied. If, for example, a return to the identities and false security of blood-and-soil nationalism is widely accepted as a foundation of political life, what chance will there be for the recognition of the rights of refugees seeking shelter from war? The forces behind the turn to nationalism, of course, involve the technologies of communication that foster discursive virulence and enclosure, but there are many other conditions—including economic ones—that inspire people in these directions. The new challenge for human rights resulting from this turn to enclosure has to do with its consequences for the leveraging effects of the exposure to public view of atrocity and injustice. What good does it do to draw attention to the crimes of the state in its handling of refugees when these "aliens" are storied as less-than-human perpetrators-in-waiting? The compliance powers of human rights extend only as far as popular conceptions of humanity will allow.

These popular conceptions have clearly shifted toward false security, enclosure, and intolerance in the new media ecosystem. With all victims of rights violations appealing to publics that are increasingly subjected to strategic campaigns of mistruth and distortion, jaded and inured to images of suffering, only those facing the clearest, most forensically verifiable, evidentially robust circumstances of illegitimate violence and mass atrocity perpetrated on the innocent will provoke a consistent, convincing public outcry. There is little room for doubt or nuance in the messages carried by shouting above the noise.

Notes

Preface

1. Zweig, 1998, *Sternstunden der Menschheit*, 153–76.

2. It is this very use of instruments of surveillance and data mining that leads me to prefer the term *information technologies* (ITs) rather than *information and communications technologies* (ICTs) throughout this book. The most significant aspects of acts of communication have become the informational purposes they serve: the repurposed data points processed by AI-driven analytics, beyond the reckoning of the media user and the particular messages they convey.

3. Zuboff, 2019, *The Age of Surveillance Capitalism*.

Introduction: Utopia and Despair

1. On extraordinary feats in charismatic authority, see Weber, 1946, *From Max Weber: Essays in Sociology*, 246.

2. Miguel Luengo Oroz, 2018, "From Big Data to Humanitarian in the Loop Algorithms," United Nations Global Pulse, 22 January 2018, https://www.unglobalpulse.org/news/big-data-humanitarian-loop-algorithms. For a report on the work of Global Pulse in the area of social media and forced displacement, see http://unglobalpulse.org/sites/default/files/White%20Paper%20Social%20Media%203_0.pdf. I am grateful to Maria Sapignoli for bringing this material to my attention.

3. UNDESA, 2018, "Frontier Technologies."

4. Giedion, [1948] 2013, *Mechanization Takes Command*, 715.

5. McLuhan, [1962] 2011, *The Gutenberg Galaxy*, 114.

6. McLuhan, [1964] 2017, *Understanding Media*, 80.

7. MacKinnon, 2012, "The Netizen."

8. MacKinnon, 2012, "The Netizen."

9. Zuboff, 2019, *The Age of Surveillance Capitalism*.

10. These figures are reported by the World Bank in a database that offers a synopsis of personal banking information. See "Spotlight: Access to Mobile Phones and the Internet around the World," 2018, *Global Findex Database 2017: Measuring Financial Inclusion*

and the Fintech Revolution (Washington, DC: International Bank for Reconstruction and Development/World Bank), 86–87, https://globalfindex.worldbank.org/sites/glo balfindex/files/chapters/2017%20Findex%20full%20report_spotlight.pdf.

11. This figure is from Kochhar, 2015, "Seven in Ten People."

12. These figures are from "Forecast Number of Mobile Users Worldwide from 2019 to 2013," Statista.com, last updated 5 July 2019, https://www.statista.com/statistics /218984/number-of-global-mobile-users-since-2010/.

13. ITU, 2017, *Measuring the Information Society Report*.

14. For a complete picture of the global growth in broadband connectivity, see ITU, 2018, *The State of Broadband 2018*.

15. In the discussion of Tarde's thought on media and publics that follows, I draw from Niezen, 2014, "Gabriel Tarde's Publics."

16. Innis, [1951] 2008, *The Bias of Communication*, 77.

17. Elias, 2001, *The Society of Individuals*, 163.

18. Tarde, 2011, *On Communication and Social Influence*, 304.

19. Tarde, 1902, *Psychologie économique*, 30–31.

20. Tarde, 2011, *On Communication and Social Influence*, 266.

21. Tarde, 2011, *On Communication and Social Influence*, 293.

22. Bautier and Cazenave, 2005, "La presse pousse-au-crime selon Tarde," 5.

23. Lippmann, 1922, *Public Opinion*, 298.

24. Besides the contributions of Jack Goody cited in note 26, there are many others who consider the social consequences of media, including, to name a few, Marshall McLuhan, [1962] 2011, *Gutenberg Galaxy*, and [1964] 2017, *Understanding Media*; Neil Postman, 1985, *Amusing Ourselves to Death*, and 1992, *Technopoly*; and Yuval Noah Harari, 2017, *Homo Deus*.

25. Postman, 1985, *Amusing Ourselves to Death*, 24.

26. Goody, 1975, *Literacy in Traditional Societies*; Goody, 1977, *The Domestication of the Savage Mind*; Goody, 1986, *The Logic of Writing and the Organization of Society*; Goody, 1987, *The Interface between the Written and the Oral*.

27. Intentionally missing from this short list of those who have tackled the question of the social influences of media technologies is Bruno Latour, particularly in the context of his actor-network theory, in which networks form the raw material out of which identities and social worlds are constructed. Latour, 2005, *Reassembling the Social*. Writing in the first years of the twenty-first century, he discusses the internet as a "technical network," together with electricity, trains, and sewage (129), giving a quasi-animistic status to these networks as "actants" with influence on social worlds, without adequately distinguishing the material networks from technologies and networks of communication that build direct connections, in Gabriel Tarde's term, "mind to mind."

28. Merry, 2006, "Anthropology and International Law," 100. Noah Weisbord, in *The Crime of Aggression*, 2019, offers a study of the development of new standards in international criminal law, informed by the ways that new technologies can incite, facilitate, and cover up mass atrocity. This study of standard setting complements the emphasis I place on popular claims making through new media as a source of human rights compliance.

29. Here I depart somewhat from Molly Land and Jay Aronson's approach to the connection between human rights and technology, which uses international human rights law and practice to orient their discussions of technological design and implementation. Land and Aronson, 2018, *New Technologies for Human Rights Law*, 3. Public conceptions of human rights are not consistently oriented in this way, and, given the central place of public opinion in processes of compliance, my goal is to convey popular understandings of human rights, whether accurately informed by international law or not.

30. Arkalgud and Partridge, 2017, *Web True.0*, 4–5.

31. Kozinets, 2010, *Netography*.

32. Coleman, 2012, *Coding Freedom*.

33. Hsu, 2014, "Digital Ethnography."

34. Gellner and Hirsch, eds., 2001, *Inside Organizations: Anthropologists at Work*, 7–8.

35. Dhiraj Murthy offers a multistranded approach to digital ethnography, advocating an array of methods that seeks a balance between physical and digital ethnographic presence. The approach I take in this book differs in the extent to which I emphasize "physical" ethnography as a foundation for understanding the uses and human consequences of new technologies. See Murthy, 2008, "Digital Ethnography."

36. Kleinman and Kleinman, 1997, "The Appeal of Experience."

Chapter 1: Street Justice

Epigraph: Macklemore & Ryan Lewis, "Buckshot," *This Unruly Mess I've Made* (Macklemore LLC, 2016).

1. These early forms of graffiti included descriptions of romantic encounters and literary commentary in ancient Greece and Rome. See Lang, 1988, *Graffiti in the Athenian Agora*; and Milnor, 2014, *Graffiti and the Literary Landscape in Roman Pompeii*. They also included the everyday devotional and votive inscriptions of prayers and memorials of lost loved ones, hopes, and fears that can still be found in traces on medieval churches. See Champion, 2015, *Medieval Graffiti*; Pritchard, 1967, *English Medieval Graffiti*.

2. See, for example, the web page of the organization Alliance Vita, 2019, "Suppression des mots père et mère," accessed 1 July 2019, https://www.alliancevita.org/2019/02/suppression-des-mots-pere-et-mere-des-formulaires-scolaires-alliance-vita-appelle-a-la-desobeissance-civile/.

3. PETA, last updated 19 January 2018, "Only Assholes Wear Fur," https://www.peta.org/blog/assholes-wear-fur/.

4. u/Gustavwind, n.d., "Notice: Animal activists downtown are bravely sticking these on the backs of people wearing fur," Reddit, archived thread, accessed 30 June 2019, https://www.reddit.com/r/toronto/comments/7or3cu/notice_animal_activists _downtown_are_bravely_s/.

5. ilovedillpickles, n.d., "I stick these on my cat all the time," comment on u/Gustavwind, "Notice: Animal activists," accessed 1 July 2019, https://www.reddit.com/r/toronto/comments/7or3cu/notice_animal_activists_downtown_are_bravely_s/. For a separate set of posts connecting the "I'm an asshole" sticker with cats, see u/cityoflostwages,

n.d., "I'm an asshole. I wear fur," Reddit, archived thread, accessed 1 July 2019, https://www.reddit.com/r/pics/comments/ca47e/im_an_asshole_i_wear_fur/.

6. An autobiographical chronicle of hip-hop's connection to graffiti in Chicago is offered in Wimsatt 2001, *Bomb the Suburbs*. See Farkas, 2012, *Toronto Graffiti* for a series of interviews with Toronto-based graffiti artists that chronicles their aspirations and activities.

7. A recent international literature offers illustrated autobiographical accounts of life in graffiti subcultures. See, for example, Mantovani, 2017, *Vecchia Scuola*; Gal, 2016, *Truskool*; and Obk, 2017, *Marqué à vie!*

8. The basic rules of the designated graffiti walls are explained in a city-sponsored web page that promotes the idea of legal walls for graffiti artists, which are indicated by the symbol of a pigeon. Wienerwand, n.d., accessed 25 August 2018, Wiener Bildungs Server, http://www.wienerwand.at.

9. *The Economist*, "This Land Is My Land," 2006.

10. "Free Fahad" (blog), 2007.

11. For information on this immigrant reception initiative, see Emmaüs Solidarité, 3 November 2016, "Le Centre Humanitaire Paris-Nord," https://www.emmaus-solidarite .org/le-centre-humanitaire-paris-nord/; and "Visitez le Centre humanitaire Paris-Nord," 16 November 2016, YouTube (video released by the office of the mayor of Paris), https://youtu.be/ObD6FY-Y8CU.

12. Cozzolino, 2017, *Peindre pour agir*, 13.

13. Crettiez and Piazza, 2014, *Murs rebelles*, 11.

14. Walter Armbrust provides a media-rich ethnographic account of the Egyptian revolution and its undoing in *Martyrs and Tricksters*, 2019.

15. Author's translation. "La publicité est l'un des vecteurs directs et violents du sexisme. Des normes sexuées sont chaque jour martelées dans les esprits. La publicité participe à la construction du genre féminin et masculin, véritables contraintes sociales imposées aux individus en fonction de leur sexe biologique." ASSÉ, 2013, Dépliant explicatif.

16. Author's translation. "netzpolitik.org ist eine Plattform für digitale Freiheitsrechte. Wir thematisieren die wichtigen Fragestellungen rund um Internet, Gesellschaft und Politik und zeigen Wege auf, wie man sich auch selbst mit Hilfe des Netzes für digitale Freiheiten und Offenheit engagieren kann. Mit netzpolitik.org beschreiben wir, wie die Politik das Internet durch Regulation verändert und wie das Netz Politik, Öffentlichkeiten und alles andere verändert. Wir verstehen uns als journalistisches Angebot, sind jedoch nicht neutral. Unsere Haltung ist: Wir engagieren uns für digitale Freiheitsrechte und ihre politische Umsetzung." Netzpolitik.com, 2015, "Über uns."

17. In subsequent visits to Berlin, I found myself in Potsdamer Platz a few times and followed up on this sticker. Even as it faded through the erosive effects of sun, rain, sleet, and snow, the URL held its form for about two years before gradually losing definition and legibility.

Chapter 2: Human Rights 3.0

1. These treaties are, taken together, those that we know today as the "bill of human rights," including the Universal Declaration of Human Rights (UDHR, adopted

in 1948), the International Covenant on Civil and Political Rights (ICCPR, adopted in 1976), and the International Covenant on Economic, Social and Cultural Rights (ICESCR, adopted in 1976).

2. Moyn, 2010, *The Last Utopia*, 3.

3. Moyn, 2010, *The Last Utopia*, 3–4.

4. I am grateful to Matthew Canfield for drawing my attention to the literature on transnational law in his remarks on a draft paper I coauthored with Maria Sapignoli, "Global Legal Institutions," for a meeting of the contributors to the *Oxford Handbook of Law and Anthropology*, held in Berlin, November 2018.

5. Zumbansen, 2008, "Transnational Law," 738.

6. Zumbansen, 2008, "Transnational Law," 742.

7. Manovich, 2001, *The Language of New Media*, 8–9.

8. McLuhan, [1964] 2017, *Understanding Media*, 38. McLuhan attributes the concept of the break boundary to Kenneth Boulding without providing source information, and then, in his inimitable way, contextualizes it historically in such a way as to make it his own.

9. Tarde, 2011, *On Communication and Social Influence*.

10. Klarsfeld and Klarsfeld, 2015, *Mémoires*.

11. Klarsfeld and Klarsfeld, 2015, *Mémoires*, 187.

12. Klarsfeld and Klarsfeld, 2015, *Mémoires*, 446.

13. The media coverage of the Eichmann trial is discussed by Lipstadt, 2011, *The Eichmann Trial*, ch. 4; and Shandler, 1997, *While America Watches*.

14. Chakravarti, 2008, "More than 'Cheap Sentimentality,'" 224.

15. Crimp, 1990, *AIDS Demo Graphics*, xx.

16. The evening news protest has since been uploaded on an ACT UP web page titled "DIVA-TV Netcast," accessed 11 October 2018, http://www.actupny.org/diva/CBnews .html.

17. Metz, 2012, "How the Queen of England."

18. Ó Muíneacháin, 2012, "Thanks, Al Gore."

19. Brodkin, 2011, "The MIME Guys."

20. Zimmerman and Emspak, 2017, "Internet History Timeline."

21. Manovich, 2001, *The Language of New Media*, 20.

22. Cleaver, 1998, "The Zapatista Effect," 622.

23. Cleaver, 1998, "The Zapatista Effect," 623.

24. A social network analysis tracked the hyperlinks connected to the Zapatistas and found that they had indeed had an impact on a significant part of the web. Garrido and Halavais, 2003, "Mapping Networks of Support."

25. For an overview of the causes, strategies, and consequences of the rebellion, see Collier and Lowery Quaratiello, 2005, *Basta! Land and the Zapatista Rebellion*.

26. I am using the word *emulate* here in two senses: the usual meaning of "imitation" but also as applied to computing: reproducing the function or action of a different computer or software system.

27. Ronfeldt and Arquilla, 2001, "Emergence and Influence of the Zapatista Social Netwar," 187.

28. Cleaver, 1995, "The Zapatistas and the Electronic Fabric of Struggle."

29. Cited in Rovira, 2009, *Zapatistas sin fronteras*, 72.

30. Rovira, 2009, *Zapatistas sin fronteras*, 74.

31. Cleaver, 1995, "The Zapatistas and the Electronic Fabric of Struggle."

32. Froehling, 1997, "The Cyberspace 'War of Ink and Internet' in Chiapas, Mexico," 296.

33. Cleaver, 1998, "The Zapatista Effect," 629.

34. Rovira, 2009, *Zapatistas sin fronteras*, ch. 3; and Cleaver, 1998, "The Zapatista Effect," 630.

35. Castells, 1997, *The Information Age*, vol. 2, *The Power of Identity*, 79.

36. Tufekci, 2017, *Twitter and Tear Gas*, 28.

37. Manovich, 2001, *The Language of New Media*, 258.

38. Niezen, 2005, "Digital Identity."

39. Obst and Stafurik, 2010, "Online We Are All Able Bodied," 530.

40. See Phillips, 2015, *This Is Why We Can't Have Nice Things*.

41. Eysenbach et al., 2004, "Health Related Virtual Communities and Electronic Support Groups," 1166.

42. Sejnowski, 2018, *The Deep Learning Revolution*, 3.

43. Zuboff, 2019, *The Age of Surveillance Capitalism*.

44. The political implications of "antisocial" media in the US context are comprehensively discussed in Benkler, Faris, and Roberts, 2018, *Network Propaganda*.

45. Ott, 2017, "The Age of Twitter," 64.

46. Horton, 2016, "Microsoft Deletes 'Teen Girl' AI."

47. Vaidhyanathan, 2018, *Anti-Social Media*, 6.

48. Vaidhyanathan, 2018, *Anti-Social Media*, 7.

49. Vaidhyanathan, 2018, *Anti-Social Media*, 5.

50. UNHCR, 2018, *Report of the Independent International Fact-Finding Mission on Myanmar*, para. 74.

51. McLaughlin, 2018, "Facebook Blocks Accounts of Facebook's Top General."

52. McIntyre, 2018, *Post-Truth*, xiv. The involvement of Cambridge Analytica in the Brexit referendum and in the 2016 US election constitute arguably the most prominent examples of social media manipulation in the service of particular power interests. See the whistle-blower account provided by Wylie, 2019, *Mindf*ck*.

53. This is the phenomenon that Maria Sapignoli, observing such state behavior in the UN Permanent Forum on Indigenous Issues, refers to as "the violence of repetition": "the form of discursive power that takes effect through reiteration." Sapignoli, 2018, *Hunting Justice*, 333.

54. G7 2017, 2017, "G7 Declaration on Responsible States Behavior in Cyberspace."

55. Schneier, 2018, *Click Here to Kill Everybody*.

56. These consumer items include objects in the realm of sexuality. "Lioness" billing itself as "the first smart vibrator for self-experimentation" syncs with smartphones to provide feedback to the user, yes, and data on the most intimate realm of human experience to tech companies as well. https://lioness.io, accessed 24 January 2019.

57. Schneier used this term in a public presentation of *Click Here to Kill Every-*

body hosted by the Berkman Klein Center for Internet and Society at the Harvard Law School, 24 September 2018.

58. Cited in Ya-Wen Lei, 2018, *The Contentious Public Sphere*, 180. I am inspired in the paragraph that follows by Lei's discussion (ch. 7) of the Chinese state's control of online discussion and dissent.

59. Lei, 2018, *The Contentious Public Sphere*, 184–85.

60. Lei, 2018, *The Contentious Public Sphere*.

61. For a general discussion of stalking and harassment, including cyberstalking, see James and MacKenzie, 2018, "Stalking and Harrassment," 172–82. Even China's proclivity toward organized harassment by online vigilantes, encouraged by some 500 million users in an enclosed media space, acts beyond control of the state. Members of the Communist Party are among those targeted, most commonly for online evidence of consumer spending. See Hatton, 2014, "China's Internet Vigilantes." The academic literature on state-sanctioned gang stalking is spare, and online discussions of the phenomenon are limited mainly to blog posts. This phenomenon nevertheless brings to mind the East German Stasi technique referred to as *Zersetzung* ("decomposition"), which involved, among other things, disseminating malicious rumors among a target's family and friends or leaving subtle signs of entry into private space. Lacking conclusive evidence of state-sanctioned gang stalking campaigns, the most I can do is make a somewhat hypothetical leap from the historically documented programs of stalking and gaslighting to the current capacities of states to engage in online surveillance, hacking, and manipulation as an almost natural outgrowth of state ambitions and capabilities. The connection between mental illness and gang stalking is emphasized in Roisin Kibert, 2016, "The Nightmarish Online World of 'Gang Stalking,'" *Vice*, 22 July 2016, https://www.vice.com/en_us/article/aeknya/the-nightmarish-online-world-of-gang-stalking. This quality of paranoia may account for the fact that its actual use by intelligence agencies is underreported.

62. Gallagher, 2018, "Google China Prototype."

63. Cited in Gallagher, 2018, "Google China Prototype."

64. Conger, 2018, "Ex-Google Employee Urges Lawmakers."

65. Sejnowski, 2018, *Deep Learning Revolution*, 3.

66. Ford, 2018, *Architects of Intelligence*, 12.

67. Raso et al., 2018, *Artificial Intelligence & Human Rights*, 18.

68. For a discussion of the social costs of these institutions of high-intensity surveillance, see Williams, 2019, "Why Everyone Should Care About Mass E-carceration."

69. Angwin et al., 2016, "Machine Bias."

70. Martin Ford in his introduction to *Architects of Intelligence*, 11–12, provides a clear introduction to the key concepts in AI, from which I have drawn in the discussion of accountability that follows.

71. Finale Doshi-Velez and Mason Kortz discuss the problem of explainability from the perspective of the uses of AI in criminal law. Doshi-Velez and Kortz, 2017, *Accountability of AI under the Law*.

72. Sejnowski, 2018, *Deep Learning Revolution*, 123.

73. Judea Pearl, a pioneer of the probability-based model of machine reasoning known as Bayesian networks, offers further reflections on the loss of transpar-

ency in deep-learning networks in Pearl, 2019, "The Limitations of Opaque Learning Machines."

74. Cited in Roberts, 2018, "The Yoda of Silicon Valley."

75. Samuels, 2019, "Is Big Tech Merging With Big Brother?"

76. Take Back the Internet with Tor (website), https://www.torproject.org/about/history/.

77. Cited in Shane and Wakabayashi, 2018, " 'The Business of War.' "

78. MoveOn, 2018, "Students Pledge to Refrain from Interviewing with Google."

79. "Celebrating War Criminals at MIT's 'Ethical' College of Computing," 14 February 2019, *The Tech*, https://thetech.com/2019/02/14/celebrating-war-criminals.

80. Harari, 2017, *Homo Deus*, 313.

81. Project Include (website), https://projectinclude.org/; see also Statt, 2016, "Ellen Pao Launches Advocacy Group."

82. Wang et al., 2019, *Gender Trends in Computer Science Authorship*.

83. Couture, 2019, "The Ambiguous Boundaries of Computer Source Code," 137.

84. I am grateful to Maria Sapignoli for conversations and materials that contributed to this subsection. More on tech-sector in UN initiatives can be found in Sapignoli and Niezen, in press, "Global Legal Institutions."

85. Zuboff, 2019, *The Age of Surveillance Capitalism*, 10.

86. Zuboff, 2019, *The Age of Surveillance Capitalism*, 11.

87. Zuboff, 2019, *The Age of Surveillance Capitalism*, 338.

88. United Nations Global Pulse (website), https://www.unglobalpulse.org. The UN presents its information technology initiatives and policies at United Nations, 2018, "The Secretary-General's Strategy on New Technologies," https://www.un.org/en/newtechnologies/images/pdf/SGs-Strategy-on-New-Technologies.pdf.

89. Microsoft, n.d., *A Digital Geneva Convention to Protect Cyberspace*.

90. Global Internet Forum to Counter Terrorism (website), https://www.gifct.org.

91. Sapignoli and Niezen, in press, "Global Legal Institutions."

92. Merry, 2016, *The Seductions of Quantification*; and Merry, in press, "Law in Practice."

93. Johns, 2016, "Global Governance."

94. The term *agility* comes from a technician's description to me of the process of developing code in the context of Canada's government administration, but it also applies in the context of global governance.

95. Doshi-Velez and Kortz, 2017, *Accountability of AI under the Law*.

96. Johns, 2016, "Global Governance," 141.

97. Johns, 2017, "Data Mining as Global Governance."

98. Johns, 2017, "Data Mining as Global Governance," 5.

99. Jigsaw, "Vision," accessed 24 January 2019, https://jigsaw.google.com/vision/.

100. Cohen and Fuller, 2013, "Fighting Human Trafficking." The link to this Google.org blog article can be found in context in the Google Jigsaw website at https://jigsaw.google.com/projects/#human-trafficking-hotline-network. See also "Palantir and Salesforce are Enabling Anti-Trafficking Organizations to Fight

Back, Thanks to Funding Help from Google," https://www.palantir.com/pt_media /how-google-is-fighting-sex-trafficking-with-big-data/.

101. Greenberg, 2016, "Google's New YouTube Analysis App."

102. Lee, 2016, "Google Targets 'The Fog of War.'"

Chapter 3: Belling the Cat

1. Aesop, "The Cat and the Bell," *Aesop for Children*, http://mythfolklore.net/aeso pica/milowinter/6.htm.

2. Freedland, 2018, "Russia's Brazen Lies Mock the World."

3. Tufekci, 2018, "It's the (Democracy-Poisoning) Golden Age."

4. Miller, 2018, "Egypt Leads the Pack."

5. "Lebanese Tourist Jailed for Facebook Post Leaves Egypt after Release," 2018, Reuters, 14 September 2018, https://www.reuters.com/article/us-egypt-politics/lebanese-tourist -jailed-for-facebook-post-leaves-egypt-after-release-idUSKCN1LT3EF.

6. Zayadin, 2018, "We Have Two Months."

7. Chen and Cheung, 2017, "The Transparent Self under Big Data Profiling."

8. Davis, in press, "'Don't Complain of the Dark.'"

9. Pils et al., 2016, "'Rule by Fear?'"

10. Davis, in press, "'Don't Complain of the Dark.'"

11. Alexander et al., 2018, *Familiar Feeling*.

12. Citizen Lab characterizes the malware in the following technical terms: "exploit-laden PowerPoint (CVE-2017-0199) and Microsoft Rich Text Format (RTF) documents (CVE-2017-11882) attached to e-mail messages. The malware includes a PowerShell payload we call DMShell++, a backdoor known as TSSL, and a post-compromise tool we call DSNGInstaller."

13. See Berkman Klein Center's Internet Monitor initiative at https://cyber.harvard.edu/research/internetmonitor, and https://thenetmonitor.org. The most promising tool the center is developing from the perspective of human rights activists, which is currently in closed beta and probably several years away from release, would let users test websites to confirm whether they are being blocked.

14. Basu, 2007, *Spy Princess*.

15. The film *Soldier of Orange* (Soldaat van Oranje) depicts the underground struggle, including the use of radio by the resistance, in the context of the Nazi occupation of Holland.

16. The Whistle.org (website), http://www.thewhistle.org.

17. "Collateral Freedom: Thwarting Censorship in 13 'Enemy of the Internet' Countries," Reporters Without Borders, https://rsf.org/en/collateral-freedom-thwarting -censorship-13-enemy-internet-countries.

18. "Front Line Defenders Named Winner of 2018 United Nations Human Rights Prize," Front Line Defenders, 25 October 2018, https://www.frontlinedefenders.org/ en/statement-report/front-line-defenders-named-winner-2018-united-nations-human -rights-prize.

19. GreatFire.org (website), 2018, https://en.greatfire.org.

20. Manovich, 2001, *The Language of New Media*, 223.

21. McPherson, 2018, "Risk and the Pluralism of Digital Human Rights Fact-Finding and Advocacy," 193.

22. eyeWitness (website), https://www.eyewitnessproject.org.

23. See, for example, Guardian Project, "CameraV: Secure Verifiable Photo and Video Camera," https://guardianproject.info/apps/camerav/. The archiving tool, Keep, developed by RightsLab, specializes in "smart, powerful archiving" of user-generated content, with newsrooms identified as likely beneficiaries (https://www.rightslab.org/keep/). YouTube's "blurring feature" tool is oriented toward protecting the identities of those depicted in videos ("How to Protect Identities with YouTube's Blurring Feature," YouTube, 25 February 2016, https://youtu.be/vBFrVlGB9Lo). The logic of digital witnessing has of course found its way into commercial applications and hardware, with advanced surveillance technologies available to public users (for example, Mobile Witness Software, http://www.mobilewitness.com/products/mobile-witness-software/). It is now possible to surround oneself with preverified, forensically authoritative digital data.

24. Land and Aronson, 2018, *New Technologies for Human Rights Law*, 400.

25. Eliot Higgins, telephone interview with the author, 28 February 2019.

26. This information is based on my conversations with Bellingcat staff at the workshop in Amsterdam, and, aside from the detail about expansion into Colombia, is confirmed by the information conveyed an article in the *New Yorker*. Beauman, 2018, "How to Conduct an Open Source Investigation." More on the methods, ethics, and potential of open source investigations can be found in Dubberley, Koenig, and Murray, eds., 2020, *Digital Witness*.

27. Eliot Higgins, telephone interview with the author, 28 February 2019.

28. *Süddeutsche Zeitung*, 2015, "Ein-Man-Nachrichtenagentur," 1 June 2015, https://archive.is/S690b.

29. "The People Who Fell from the Sky," Bellingcat, podcast, 17 July 2019, https://www.bellingcat.com/resources/podcasts/2019/07/17/mh17-episode-guide-1/), 2:28.

30. Bellingcat Investigation Team, 2018, "Anatoliy Chepiga."

31. Bellingcat Investigation Team, 2018, "Full Report: Skripal Poisoning Suspect."

32. Bellingcat Investigation Team, 2019, "The GRU Globetrotters."

33. *RT News*, 2018, "'Suck My Balls.'"

34. Bellingcat Investigation Team, 2018, "Chemical Weapons and Absurdity."

35. Eliot Higgins, telephone interview with the author, 28 February 2019.

36. Pieter van Huis, Bellingcat workshop, Amsterdam, 4 February 2019.

37. Rawan Shaif, Twitter, https://twitter.com/RawanSSA18.

38. Higgins, 2017, *Op zoek naar die waarheid*, 7.

39. The Bellingcat Foundation Policy Plan describes its central ambition in obtaining NGO status in The Netherlands: "to become the global hub for training and research and . . . to centralize (and professionalize) all activities run by Bellingcat internationally." https://www.bellingcat.com/wp-content/uploads/2019/09/ANBI-beleidsplan.pdf.

40. Christiaan Triebert, from his position as an investigative journalist for the *New York Times*, continues to support the work of Bellingcat through active participation in its network of investigators, with "previously @bellingcat" part of his Twitter profile.

41. See, for example, Daniel Romein, 2019, "More Europol's 'Stop Child Abuse' Photographs Geolocated," Bellingcat, 2 July 2019, https://www.bellingcat.com/news/uk-and -europe/2019/07/02/more-europols-stop-child-abuse-photographs-geolocated/.

42. Syrian Archive (website), https://syrianarchive.org/en.

43. Atlantic Council Digital Forensic Research Lab, "About," https://www.digital sherlocks.org/about.

44. #digitalsherlocks, Twitter, https://twitter.com/search?q=%23digitalsherlocks&src =tyah&lang=fr.

45. Atlantic Council Digital Forensic Research Lab, "About," https://www.digital sherlocks.org/about.

46. SITU Research (website), https://situ.nyc/research/people.

47. Interview with Brad Samuels, SITU Research, Brooklyn, 6 November 2019. The approach to forensics taken by Forensic Architecture as a "critical examination of the prevalent status of forensics in articulating contemporary notions of public truth" is discussed in Weizman, 2014, "Introduction: Forensis."

48. Goldsmiths' Center for Research Architecture and SITU Studio, 2009, "Report."

49. SITU Research, 2013, *Bil'In Report*, https://situ.nyc/research/projects/bilin -report.

50. Forensic Architecture (website), https://www.forensic-architecture.org.

51. Halbfinger, 2018, "A Day, a Life."

52. Ruser, 2018, *How to Scrape Interactive Geospatial Data.*

53. Strick, 2018, *How to Identify Burnt Villages.*

54. Toler, 2018, *Creating an Android Open Source Research Device on Your PC.*

55. "Violent or Graphic Content Policies," YouTube Help, accessed 24 February 2019, https://support.google.com/youtube/answer/2802008?hl=en.

56. Rawan Shaif, Twitter, https://twitter.com/RawanSSA18.

57. The term *hackathon* no longer necessarily carries connotations of hacking, or overcoming the limitations of software systems, often to exploit weaknesses in computer security.

58. Human Rights Center, "Open Source Investigations Protocol," UC Berkeley–Berkeley Law, accessed 28 June 2019, https://humanrights.berkeley.edu/programs-projects /tech-human-rights-program/open-source-investigations-protocol.

59. SITU Research, "ICC Digital Platform: Timbuktu, Mali," https://situ.nyc/research /projects/icc-digital-platform-timbuktu-mali.

60. ICC, 2017, "Situation in Libya," Warrant of Arrest, 15.

61. BBC World Service, Twitter thread, "War Crimes for Likes," 2 May 2019, https:// twitter.com/bbcworldservice/status/1123895595863822338.

62. See Bellingcat Investigation Team, 2017, "How a Werfalli Execution Site Was Geolocated," Bellingcat, 3 October 2017, https://www.bellingcat.com/news/mena/2017 /10/03/how-an-execution-site-was-geolocated/.

63. ICC, 2018, "Situation in Libya," Second Warrant of Arrest, para 18.

64. Osman, 2018, "ICC Suspect Al-Werfalli 'Escapes.' "

65. Beauman, 2018, "How to Conduct an Open Source Investigation."

Chapter 4: Shouting Above the Noise

1. Cited in Sarogni, 2004, *La Donna Italiana*, 39–40). My translation.

2. *La Stampa*, 2015, "Umberto Eco, 'Con i social parola a legioni di imbecilli.'"

3. Berger, 2016, *Contagious: Why Things Catch On*, 15.

4. Boo, 2007, "Difficult Journalism That's Slap-Up Fun," 14.

5. Roberts and Stalans, 2000, *Public Opinion, Crime, and Criminal Justice*, 13–14.

6. Giorgio Resta, 2010, "Il problema dei processi mediatici nella prospettiva del diritto comparato," 21.

7. "Tweets about Global Development Topics," UN Global Pulse, accessed 27 February 2019, http://post2015.unglobalpulse.net.

8. Niezen, 2010, *Public Justice and the Anthropology of Law*, ch. 2.

9. UNHCR, "Horn of Somalia Situation," Operational Portal: Refugee Situations, accessed 8 September 2018, https://data2.unhcr.org/en/situations/horn?id=3&country =110.

10. Federman, 2018, "The 'Ideal Perpetrator.'"

11. Federman, 2018, "The 'Ideal Perpetrator,'" 330.

12. Eubanks, 2017, *Automating Inequality*, 9.

13. Asal and Harwood, 2008, "Airing Grievances Online," 4.

14. Brin and Page, 1998, "The Anatomy," 109.

15. Noble, 2018, *Algorithms of Oppression*, 3.

16. Noble, 2018, *Algorithms of Oppression*, 42–43.

17. O'Neil, 2017, *Weapons of Math Destruction*, 228.

18. Browning, 2018, "The Suffocation of Democracy."

19. Merry, 2016, *The Seductions of Quantification*, 1.

20. Merry, in press, "Law in Practice."

21. I am grateful to Sarah Federman for sharing with me her decade of experience in a Manhattan-based advertising firm as the basis of this foregoing analysis (27 November 2018).

22. *The Guardian*, 2017, "Environmental Defenders Being Killed in Record Numbers Globally, New Research Reveals," 13 July 2017, https://www.theguardian.com/environment/2017/jul/13/environmental-defenders-being-killed-in-record-numbers-globally-new-research-reveals.

23. Small, Loewenstein, and Slovic, 2007, "Sympathy and Callousness."

24. Austin, 2006, "The Next 'New Wave.'"

25. Woodside, Sood, and Miller, 2008, "When Consumers and Brands Talk," 100.

26. Kristof, 2018, "Be Outraged by America's Role in Yemen's Misery."

27. Kang, 2018, "Executives Pull out of Saudi Conference."

28. Marrus, 2002, *The Unwanted*, 51–52.

29. Digital Smoke Signals (website), http://www.digitalsmokesignals.com.

30. Myron Dewey, 2018, presentation for Civic Arts Series, Comparative Media Studies/Writing, Massachusetts Institute of Technology, Cambridge, MA, 15 November 2018.

31. TigerSwan: Solutions to Uncertainty (website), https://www.tigerswan.com.

32. The concept of visual legal narrative, used by the documentary filmmaker

Myron Dewey (personal communication, November 2018) is discussed in Austin, 2006, "The Next 'New Wave,'" under the rubric Visual Legal Advocacy.

33. *Awake: A Dream from Standing Rock* (website for film), https://awakethefilm.org.

34. Becker, 2013, *Campaigning for Justice*, 248–49.

35. Tate, 2013, "Proxy Citizenship and Transnational Advocacy," 58.

36. For a more complete account of the practices of *testimonio* by Colombian activists, see the ethnographic study, Tate, 2007, *Counting the Dead.*

Chapter 5: Media War

1. In Tamashek legend, Imollen is a blind visionary, accompanied almost everywhere he goes by the forger Awjembak.

2. Translated from the Tamashek by Hélène Claudot-Hawad and translated from the French by me. (Not ideal, but the French translation is excellent and done in collaboration with Hawad, who is fluent in both languages.)

3. As part of the research for this book, I attended the 2018 and 2019 meetings of the Tuareg Diaspora in Europe (Rencontre de la Diaspora Touaregue en Europe) in the village of Le Cormier in Normandy, France, 14 and 15 July 2018 and 13 and 14 July 2019. The programs of the meeting are posted online: http://www.tamoudre.org/touaregs/musique/rencontre-annuelle-de-la-diaspora-en-europe-14-15-juillet-2018-le-cormier-normandie/; and http://www.tamoudre.org/touaregs/societe/diaspora-touaregue-en-europe-rencontres-2019-13-14-juillet-2019-haute-normandie/.

4. Keenan, 2004, *The Lesser Gods of the Sahara*, 56.

5. Although I have not visited Gao since the construction of the peacekeeping mission, the site can be visited online, including in a time-lapse of its construction, on Google Earth Pro, one of the key tools of crowdsourced human rights forensics.

6. The sporadic killing of peacekeepers in Gao is occasionally covered in the mainstream media, but—possibly because it is intermittent and involves mainly African peacekeepers, remote from the interest and attention of news consumers in the West—it has not received wide coverage. For a comparative overview of the violence targeted at peacekeepers that remains valid, see *BBC News*, 2015, "World's Most Dangerous Peacekeeping Mission."

7. I discuss this Islamic reform movement at greater length elsewhere but with a focus on the Songhay rather than the Tuaregs, who took up Salafi-inspired Islamic reform several decades later. Niezen, 1990, "The Community of Helpers of the Sunna"; and Niezen, 1991, "Hot Literacy in Cold Societies." On the temporary alliance between the MNLA and Ansar Dine, see *BBC News*, 2012, "Mali, Tuareg and Islamist Rebels Agree on Islamist State."

8. The most basic division frequently noted in the literature on Tuareg society is the binary structure of nobles and vassals, the classification that distinguishes the *imuhagh* (raiders) and the *kel ulli* (people of the goat). This dualism is based on an oversimplified distinction between subsistence strategies based on long-range camel herding in the desert regions and shorter-range goat herding, limited mostly to the more abundant savannah further south. Historically, the *imuhagh* noble lineages that wielded the most political

power were those who possessed the greatest number of camels and who traveled furthest in search of viable pasture, and who therefore also possessed the means of mobility and desert warfare. The *kel ulli*, by contrast, did not travel as far and remained more closely tied to settled life. There are other, more sophisticated breakdowns of Tuareg social structure, which add other strata or "castes" to this basic dualism, one of which identifies five basic groups: the *imazighen*, or nobility; the *ineslimen*, or clerical class; the *imghad*, or vassals; the *inaden*, or handicraft specialists; and the *iklan*, or former captives or servants.

    9.   Kohl and Fischer, 2010, "Tuareg Moving Global," 4.

   10.   The trans-Saharan trade, once a mainstay of commerce among camel-herding nomads, was interrupted by changes in the global economy before being subjected to direct interference by the French and British.

   11.   Kohl and Fischer, 2010, "Tuareg Moving Global," 5.

   12.   Morgan, 2012, "The Causes of the Uprising in Northern Mali."

   13.   Baqué, 1995, "Nouvel enlisement des espoirs de paix," 30.

   14.   Klute, 2010, "Kleinkrieg in der Wuste."

   15.   Rasmussen, 2006, *Spirit Possession and Personhood.* 16. I do not have space here to enter into the place of women in Tuareg descent systems and access to leadership, which have been largely disrupted through the contemporary shifts in values and conflicts that are my focus here. Dida Badi builds on this missing feature in descriptions of Tuareg sociopolitical organization to make the point that the institution of the lineage leader, or *amenukal*, has its origin in the female descent practiced by the sedentary Tuareg, in which women occupy a privileged place as the source of historical depth, rootedness to the land, and access to political power. Badi, 2010, "Genesis and Change in the Socio-Political Structure of the Tuareg," 80.

   16.   See Pandolfi, 2001, "Les touaregs et nous."

   17.   Lunacek, 2010, "Ambiguous Meanings of *Ikufar*," 197.

   18.   Mériadec, 2013, *Touaregs: La révolte des hommes bleues.*

   19.   As Didi Badi points out, the simple lineage model is incomplete and largely out of date, representing the Tuareg social order in the particular ecological niche of desert pastoralism during the late colonial period, with a focus that captured only "one step . . . in a precise moment in time in this society's long advance through desert space." Badi, 2010, "Genesis and Change in the Socio-Political Structure of the Tuareg," 79.

   20.   Baryin, 2013, *Dans les mâchoires du chacal,* 42.

   21.   "Members of the Permanent Forum," United Nations Department of Economic and Social Affairs, Indigenous Peoples, accessed 27 January 2019, https://www.un.org/development/desa/indigenouspeoples/unpfii-sessions-2/newmembers.html.

   22.   Falk, 2001, Preface, in *The Right to Self-Determination,* 6.

   23.   Morgan, 2012, "Causes of the Uprising in Northern Mali."

   24.   Bosman, 2013, "Mali," 252–53.

   25.   Morgan, 2012, "Causes of the Uprising in Northern Mali."

   26.   *The National,* 2013, "Tuareg Insurgents in Mali," Bloomberg News.

27. Bosman, 2013, "Mali," 356.

28. Lecocq and Klute, 2013, "Tuareg Separatism in Mali," 431.

29. Keenan, 2012, "How Washington Helped Foster the Islamist Uprising in Mali."

30. Albachir, 2102, "Déclaration de l'association TUNFA."

31. Aboubacrine, 2012, "Alerte sur la question des peoples autochtones."

32. African Union, 2015, "Communiqué of the Peace and Security Council."

Chapter 6: The Politics of Memory

Epigraph: Hannah Arendt, *Crises of the Republic* (San Diego: Harcourt Brace, 1972). This chapter is derived in part from an article published in the *International Journal of Heritage Studies*, 22 December 2017, copyright Taylor & Francis, available online: http://www.tandfonline/10.1080/13527258.2017.1413681.

1. In this chapter, I use the term *Ovaherero* in preference to *Herero*. The *ova-* prefix in Bantu languages commonly refers to men in general or "men spread over the earth," which makes the same distinction as that between the *Herero* and the *Herero people*. *Ovaherero* is therefore the noun form and *Herero* the adjectival form of the ethnonym.

2. Not surprisingly, the numbers of those killed in the colonial war are contested. A thorough discussion of the conflict as genocide, including the numbers killed, is offered in Wallace, 2011, *A History of Namibia*.

3. The stone reads: "To the 41 members of the regiment who participated voluntarily in the period from January 1904 to March 1907 on campaigns in Southwest Africa and died the hero's death. The officer corps honors the memory of the heroes with this stone." The stone marks the end of the conflict as 1907, and there is similar ambiguity in the literature about the dates of the war, with some authors using the years 1904–1907 and others marking the end of the conflict in 1908. Because, in my view, the continuation of ethnic extermination marks a continuation of the war, in this chapter I will use the year 1908 to mark the end of the conflict.

4. "Zum Gedenken an die Opfer der deutschen Kolonialherrschaft in Namibia 1884–1915 insbesondere des Kolonialkrieges von 1904–1907. Die Bezirksverordnetenversammlung und das Bezirksamt Neukölln von Berlin. Nur wer die Vergangenheit kennt hat eine Zukunft—Wilhelm von Humboldt." Author's translation.

5. This question can be approached differently depending on the sense of "knowing" one follows. That used by Humboldt, writing in the early nineteenth century, is derived from the verb *kennen*, to be familiar with, rather than the concept of *Verstehen* later developed by Max Weber, meaning the interpretive or participatory understanding of social phenomena.

6. Kössler, 2007, "Facing a Fragmented Past," 365.

7. Kössler, 2007, "Facing a Fragmented Past," 364–65.

8. "Unser Ziel ist es, das zivilgesellschaftliche Engagement für Demokratie und Toleranz in unserem Land sichtbar zu machen." Bündnis für Demokratie und Toleranz, "Über uns," accessed 26 August 2017, http://www.buendnis-toleranz.de/ueberuns/.

9. Afrotak TV, 2016, "No Amnesty on Genocide Deutschland," 24 August 2016, http://afrotak.com/2016/08/24/no-amnesty-on-genocide-deutschland/.

10. "Das Projekt bringt den 1. Völkermord der Moderne im heutige Namibia (1904–1908) durch Deutschland in den öffentlichen Raum ein, der als Vorläufer des NS-Holocausts in Deutschland prägend auf den Umgang mit 'Den Anderen' gewirkt hat." Afrotak TV, "No Amnesty on Genocide Deutschland," accessed 25 August 2017, http://afrotak.com/2016/08/24/no-amnesty-on-genocide-deutschland/.

11. The Ovambanderu (or Mbanderu) are an Otjiherero-speaking people of eastern Namibia and western Botswana, whose history of subjection to colonial rule and genocide and the contemporary claims that go with it parallel that of the main body of the Ovaherero in Namibia. For purposes of simplicity, I include them in the ethnonym Ovaherero, in keeping with the common practice in the literature.

12. Smith, 1974, *Corporations and Society*, 101.

13. The same point could be made for the Tuaregs, as discussed in the previous chapter, but in their case the formation of corporate groups has inclined more toward fragmentation around distinct leadership and political values.

14. See, for example, the website of Freiburg-Postkolonial, http://www.freiburg -postkolonial.de/Seiten/2009-Zeller-Namibiagedenkstein-Berlin.htm.

15. Merry, 2006, *Human Rights and Gender Violence*, 3.

16. de Gaay Fortman, 2011, *Political Economy of Human Rights*. The idea of "up-streaming" is in some ways consistent with Gramsci's use of the concept of hegemony. Contrary to the usual notions of hegemony as the imposition and passive acceptance of state-sanctioned ideologies, for Gramsci the subaltern classes were not empty vessels but actively engaged with knowledge and political acts of communication. Gramsci, [1929-1932] 1975, *Quaderni del carcere*.

17. This observation was made at a public symposium on the Herero/Nama genocide sponsored by the Grassi Museum für Völkerkunde zu Leipzig, 13 December 2018.

18. See, for example, Nora Krug's best-selling autobiographical graphic novel *Heimat: Ein deutsches Familienalbum* (2018), which movingly engages with the youth generation's struggles to find meaning in the context of morally complex intersections of family histories and world history.

19. Government spokesperson Martin Schäfer expressed the government's and parliament's recognition of the genocide in the following terms: "Der Vernichtungskrieg in Namibia von 1904 bis 1908 war ein Kriegsverbrechen und Völkermord" (The war of elimination in Namibia from 1904 to 1908 was a war crime and genocide). Deutsche Welle, 2015, "Bundesregierung nennt Namibia-Massaker 'Völkermord.'"

20. Melber and Kössler, 2015, "Wer B sagt, muss auch A sagen."

21. Lu, 2017, *Justice and Reconciliation in World Politics*; and Steinhauser, 2017, "Germany Confronts the Forgotten Story."

22. Kössler, 2015, *Namibia and Germany*. Esther Muinjangue offered a similar view in a press conference in Berlin on 12 October 2017: "If you acknowledge that it is genocide, on which basis do you decide that there is no reparation?"

23. Such views are supported by Kössler's historical perspective, which finds that the genocide committed in Namibia and the colonial violence of Tanzania have been "expunged from national memory." Kössler, 2007, "Facing a Fragmented Past," 364–65.

24. In the words of founding president Sam Nujoma, "We are leaving the sad history behind us and working progressively together" (Wir lassen die traurige Geschichte hinter uns zurück und arbeiten fortschrittlich zusammen) *Die Welt*, 2002, "Hört endlich auf mit eurer weißen Arroganz." All translations are by the author.

25. Because of incomplete records, the 80 percent figure is of course an approximation. Olusoga and Erichsen put the death rate of the Nama in captivity higher, at 90 percent. Olusoga and Erichsen, 2011, *The Kaiser's Holocaust*, 229. Accounts of camp conditions and death tolls are also provided by Hull and Kuss. Hull, 2005, *Absolute Destruction*, 88–90; and Kuss, 2004, "Der Herero-Deutsche Krieg."

26. Hull, 2005, *Absolute Destruction*.

27. In this chapter I occasionally make use of the common shorthand "Ovaerero-Nama genocide," mainly for reasons of expediency. There is a tendency to refer to this genocide in strictly ethnic terms, which tends to exclude the less concentrated, but ultimately catastrophic, victimization of other groups such as the Damara and San.

28. Korster, 2013, "Mau Mau Reparations."

29. Crapanzano, 2011, *The Harkis*.

30. Vansina, 2010, *Being Colonized*; and Hochschild, 2012, *King Leopold's Ghost*.

31. Jeremy Sarkin and Carly Fowler offer an analysis of the first effort by the Ovaherero and Nama claimants and their lawyers to sue Germany and German corporations for the mass killing in German Southwest Africa, under the Alien Torts Claims Act. Their argument relies on evidence of violations of international law that were already in force in 1904, since the Genocide Convention cannot be applied retroactively. Sarkin and Fowler, 2008, "Reparations for Historical Human Rights Violations."

32. Bargueño, 2012, "Cash for Genocide?," 402.

33. My research for this section and the one that follows took place in Namibia from May to September 2015 and in Berlin, Germany from May to September 2017 and December 2018. My method consisted of visiting and recording the exhibits and monuments relating to state-sanctioned identity as well as the dissenting material expressions of grievance. I also met with and interviewed activists and lawyers involved in the Herero and Nama claims and explored their online expressions of their purpose and methods on Facebook and their interconnected network of websites.

34. Borneman, 1992, *Belonging in the Two Berlins*.

35. Becker, 2013, *Campaigning for Justice*.

36. Eicker, 2009, *Der Deutsch-Herero-Krieg und das Völkerrecht*, 497.

37. Irle, 1906, *Die Herero*, vi.

38. Carrier, 2005, *Holocaust Monuments and National Memory Cultures*.

39. See Young, 1992, "The Counter-Monument."

40. *Der Spiegel* online, 2005, "Remembering the Holocaust."

41. *The Telegraph*, 2017, "Chinese Tourists Arrested."

42. http://www.stolpersteine.eu/en/home/. The information included on the *Stolpersteine* can vary according to circumstances. For example, "here lived" can be replaced by "here worked," "here practiced," "here studied," and so on, and the stones can com-

memorate suicide (*Flucht in den Tod*), liberation (*Befreit*), or "fate unknown" (*Schicksal unbekannt*), all following carefully selected wording.

43. Habermas, 1988, "Concerning the Public Use of History," 44.
44. Young, 1992, "The Counter-Monument," 269.
45. Gustav Rust (website), https://www.gustav-rust-berlin.de.
46. Gensburger and Lefranc, 2017, *À quoi servent les politiques de mémoire?*, 8.
47. Kubal, 2008, *Cultural Movements and Collective Memory.*
48. Johnson, 1995, "Cast in Stone."
49. Bosch, 2017, "Twitter Activism and Youth in South Africa."
50. Volkan, 2001, "Transgenerational Transmission and Chosen Traumas."
51. Rottenburg, 2009, *Far-Fetched Facts*, xxxi. Seen another way, the Ovaherero-Nama claims represent a multistranded, technologically enabled, transnational form of postcolonial struggles with and against state power, particularly against dominant forms of the "rule of law," as described by Jean and John Comaroff. Comaroff and Comaroff, 2006, introduction to *Law and Disorder in the Postcolony.* I am grateful to George Meiu for contributing to this insight (personal communication, 25 March 2019).
52. Press conference organized by Berlin Postkolonial, Hotel de France, Berlin, 13 September 2017.
53. Kössler, 2007, "Facing a Fragmented Past," 365.
54. Savage, 2009, *Monument Wars.*
55. Habermas, 1988, "Concerning the Public Use of History," 47.
56. See Ashplant, Dawson and Roper, 2000, *The Politics of War Memory and Commemoration.*
57. Trouillot, 1995, *Silencing the Past*, xix.
58. Kössler, 2007, "Facing a Fragmented Past," 366.
59. Smith, 2006, *Uses of Heritage.*
60. Zuern, 2017, "Namibia's Monuments to Genocide."
61. I am using the term *counter-monument* in a more literal way than that invoked by Young, for whom the postwar period in Germany was marked by a wider movement toward shaping the symbols of public space. In what I observed, there was also a more focused effort to undo the symbolic effects of specific monuments by erecting dissenting monuments in adjacent space.
62. Schildkrout, 1995, "Museums and Nationalism in Namibia."
63. "Bündnis 'Völkermord verjährt nicht!,'" 2016, press release, ISD-Bund e.V., 13 June 2016, http://isdonline.de/buendnis-voelkermord-verjaehrt-nicht/.
64. "Die langjährige Weigerung aufeinanderfolgender Bundesregierungen, den Völkermord in Namibia anzuerkennen bleibt ein Armutszeugnis für die Bundesrepublik Deutschland und passt in keiner Weise zu deren Selbstbild als 'Erinnerungsweltmeister' und dem damit verbundenen beispielgebend kritischen Umgang mit der eigenen Geschichte." ISD-Bund e.V., 2016, "Bündnis 'Völkermord verjährt nicht!'"
65. See Förster et al., 2016, "Negotiating German Colonial Heritage."
66. Hägler, 2006, "Manche kämpfen weiter gegen die Hereros."

67. Lackmann, 2015, "Warum heißt die Mohrenstraße Mohrenstraße?"

68. See, in particular the Naturfreunde Jugend Berlin (Nature-Friends Youth Berlin) website, https://naturfreundejugend-berlin.de/kampagnen/pink-rabbit.

69. Africavenir, n.d., "German Genocide in Namibia," http://www.africavenir.org/advocacy/german-genocide-in-namibia.html.

70. Schildkrout, 1995, "Museums and Nationalism in Namibia," 66.

71. Kössler, 2007, "Facing a Fragmented Past," 369–70.

72. Melber, 2007, "Poverty, Politics, Power and Privilege"; and Kisting, 2017, "Namibia's Wealth Redistribution Plan."

73. Werner, 1993, "A Brief History of Land Dispossession in Namibia," 139–40.

74. Zuern, 2012, "Memorial Politics," 497.

75. du Pisani, 2010, "The Discursive Limits of SWAPO's Dominant Discourses," 2.

76. Kössler, 2007, "Facing a Fragmented Past," 362.

77. Namibia Tourism Board, "Celebrating the Heroes of Namibia," Namibia: Endless Horizons, http://www.namibiatourism.com.na/blog/Celebrating-the-Heroes-of-Namibia.

78. "Das neue Namibia und das neue Deutschland beschäftigen sich nicht mehr mit der Vergangenheit. Wir lassen die traurige Geschichte hinter uns zurück und arbeiten fortschrittlich zusammen." *Die Welt*, 2002, "Hört endlich auf mit eurer weissen Arroganz."

79. Joachim Zeller, "(Post-)Koloniale Monumente: Denkmalinitiativen erinnern an die imperiale Übersee-Expansion Deutschlands," Denkmal, Afrika-Hamburg.de, accessed 30 August 2017, http://afrika-hamburg.de/denkmal5.html. The plaque later provoked the ire of right-wing extremists, who defaced it with paint in the black, red, and white colors of the German state flag of 1871 and of the National Socialists.

80. Interview in Berlin, 12 October 2017.

81. Kocka, 1990, "Deutsche Identität und historischer Vergleich."

82. Nordbruch, 2006, *Völkermord an den Herero in Deutsch-Südwestafrika?*, 106.

83. Schildkrout, 1995, "Museums and Nationalism in Namibia," 75.

84. Kössler, 2007, "Facing a Fragmented Past," 378.

85. An account of the recent histories of the Reiterdenkmal and the Marine Denkmal is offered in Zuern, 2017, "Namibia's Monuments to Genocide."

86. Werner, 1993, "A Brief History of Land Dispossession," 139.

87. Kössler, 2007, "Facing a Fragmented Past," 368.

88. Cited in Erichsen, 2008, *"What the Elders Used to Say,"* 59.

89. Erichsen, 2008, *"What the Elders Used to Say,"* 48.

90. Cited in Steinhauser, 2017, "Germany Confronts the Forgotten Story."

91. Steinhauser, 2017, "Germany Confronts the Forgotten Story."

92. Förster, 2008, "From 'General Field Marshall' to 'Miss Genocide,'" 177.

93. Kössler, 2007, "Facing a Fragmented Past," 376.

94. For an online version of this discussion, see "Reiterdenkmal, Windhoek," Revolvy, https://www.revolvy.com/main/index.php?s=Reiterdenkmal,%20Windhoek.

95. Förster, 2008, "From 'General Field Marshall' to 'Miss Genocide,'" 188–89.

96. "German Genocide in Namibia," n.d., Advocacy, Africavenir, http://www.africavenir.org/advocacy/german-genocide-in-namibia.html.

97. Interview with Esther Muinjangue, Berlin, 12 September 2017.

98. Gross, 2018, "The Troubling Origins of the Skeletons in a New York Museum."

99. Talita Uinuses, Grassi Museum für Völkerkunde zu Leipzig, 13 December 2018.

Conclusion: Truth and Power

1. See, for example, Ott, 2017, "The Age of Twitter"; and Vaidhyanathan, 2018, *Anti-Social Media*.

2. Pariser, 2011, *The Filter Bubble*; and Sunstein, 2017, *#Republic: Divided Democracy in the Age of Social Media*.

3. de Gaay Fortman, 2011, *Political Economy of Human Rights*; and Sapignoli, 2018, *Hunting Justice*, 164–65.

4. McLuhan, [1962] 2011, *The Gutenberg Galaxy*, xli.

5. Lamberti, 2012, *Marshall McLuhan's Mosaic*, 63.

6. Costanza-Chock, 2014, *Out of the Shadows, Into the Streets!*

References

Aboubacrine, Saoudata. 2012. "Alerte sur la question des peoples autochtones et communauté locales du Nord Mali ou Azawad." Eleventh session of the Permanent Forum on Indigenous Issues, New York, 7–18 May 2012.

African Union. 2015. "Communiqué of the Peace and Security Council of the African Union at Its 544th Meeting on the Situation in Mali and the Sahel Region in General." Addis Ababa, Ethiopia, 18 September 2015. http://www.peaceau.org/en/article/communique-of-the-peace-and-security-council-of-the-african-union-au-at-its-544th-meeting-on-the-situation-in-mali-and-the-sahel-region-in-general.

Adler, Patricia, and Peter Adler. 2008. "The Cyber Worlds of Self-Injurers: Deviant Communities, Relationships, and Selves." *Symbolic Interaction* 31, no. 1: 33–56.

Albachir, Aboubacar. 2012. "Déclaration de l'association TUNFA." Eleventh session of the Permanent Forum on Indigenous Issues, New York, 7–18 May.

Alexander, Geoffrey, Matt Brooks, Masashi Crete-Nishihata, Etienne Maynier, John Scott-Railton, and Ron Deibert. 2018. *Familiar Feeling: A Malware Campaign Targeting the Tibetan Diaspora Resurfaces*. Online report, 8 August 2018. Citizen Lab, Munk School of Global Affairs, University of Toronto. https://citizenlab.ca/2018/08/familiar-feeling-a-malware-campaign-targeting-the-tibetan-diaspora-resurfaces/.

Amnesty International. 1991. *Amnesty International Report 1991*. London: Amnesty International Publications.

Angwin, Julia, Jeff Larson, Surya Mattu, and Lauren Kirchner. 2016. "Machine Bias." *ProPublica*, 23 May 2016. https://www.propublica.org/article/machine-bias-risk-assessments-in-criminal-sentencing, accessed 28 December 2018.

Arkalgud, Ujwal, and Jason Partridge. 2017. *Web True.o: Why the Internet and Digital Ethnography Hold the Key to Answering the Questions That Traditional Research Just Can't*. Morrisville, NC: Lulu.

Armbrust, Walter. 2019. *Martyrs and Tricksters: An Ethnography of the Egyptian Revolution*. Princeton, NJ: Princeton University Press.

Asal, Victor, and Paul Harwood. 2008. "Airing Grievances Online: Search Engine Algorithms and the Fate of Minorities at Risk." *Journal of Information Technology & Politics* 4, no. 3: 3–17.

Ashplant, T. G., G. Dawson, and M. Roper, eds. 2000. *The Politics of War Memory and Commemoration*. London: Routledge.

ASSÉ. 2013. *Dépliant explicatif: Sales pubs sexistes*. Association pour une solidarité syndicale étudiante, 17 March 2013. https://nouveau.asse-solidarite.qc.ca/wp-content/uploads/2013/03/depliant-sales-pubs-sexistes-2010.pdf.

Austin, Regina. 2006. "The Next 'New Wave': Law-Genre Documentaries, Lawyering in Support of the Creative Process, and Visual Legal Advocacy." *Fordham Intellectual Property, Media, and Entertainment Law Journal* 16, no. 3: 809–68.

Autry, R. 2013. "Doing Memory in Public: Postapartheid Memorial Space as an Activist Project." In *Memory and Post-War Memorials: Confronting the Violence of the Past*, edited by M. Silberman and F. Vatan, 137–54, New York: Palgrave Macmillan.

Badi, Dida. 2010. "Genesis and Change in the Socio-Political Structure of the Tuareg." In *Tuareg Society within a Globalized World: Saharan Life in Transition*, edited by Anja Fischer and Ines Kohl. New York: Tauris Academic Studies.

Baker, Keith. 1990. *Inventing the French Revolution*. Cambridge, UK: Cambridge University Press.

Ball, Patrick, Herbert Spirer, and Louise Spirer, eds. 2000. *Making the Case: Investigating Large Scale Human Rights Violations Using Information Systems and Data Analysis*. Washington, DC: American Association for the Advancement of Science.

Baqué, Philippe. 1995. "Nouvel enlisement des espoirs de paix dans le conflit touareg au Mali." *Le Monde Diplomatique*, April, 30–31.

Bargueño, D. 2012. "Cash for Genocide? The Politics of Memory in the Herero Case for Reparations." *Holocaust and Genocide Studies* 26, no. 3: 394–424.

Baryin, Gael. 2013. *Dans les mâchoires du chacal: Mes amis Touaregs en guerre au nord-Mali*. Neuvy-en-Champagne: Le passager clandestin.

Basu, Shrabani. 2007. *Spy Princess: The Life of Noor Inayat Khan*. New Lebanon, NY: Omega.

Bautier, Roger, and Élisabeth Cazenave. 2005. "La presse pousse-au-crime selon Tarde et ses contemporains" (The push-to-crime press according to Tarde and his contemporaries). *Champ pénal/Penal Field, nouvelle revue internationale de criminologie* (online). 34ème Congrès français de criminologie (34th French Conference on Criminology). http://champpenal.revues.org/253.

BBC News. 2012. "Mali Tuareg and Islamist Rebels Agree on Islamist State." 27 May 2012. http://www.bbc.co.uk/news/world-africa-18224004.

BBC News. 2015. "World's Most Dangerous Peacekeeping Mission." 20 November 2015. https://www.bbc.com/news/world-africa-34812600.

Beauman, Ned. 2018. "How to Conduct an Open Source Investigation, According to the Founder of Bellingcat." *New Yorker*, 30 August 2018. https://www.newyorker.com/culture/culture-desk/how-to-conduct-an-open-source-investigation-according-to-the-founder-of-bellingcat.

Becker, Jo. 2013. *Campaigning for Justice: Human Rights Advocacy in Practice*. Stanford, CA: Stanford University Press.

———. 2017. *Campaigning for Children: Strategies for Advancing Children's Rights*. Stanford, CA: Stanford University Press.

Beer, David. 2009. "Power through the Algorithm? Participatory Web Cultures and the Technological Unconscious." *New Media and Society* 11, no. 6: 985–1002.

Bellingcat. 2019. "The People Who Fell from the Sky," podcast. 17 July 2019. https://www .bellingcat.com/resources/podcasts/2019/07/17/mh17–episode-guide-1/, accessed 24 July 2017.

Bellingcat Investigation Team. 2018. "Anatoliy Chepiga Is a Hero of Russia: The Writing Is on the Wall." Bellingcat, 2 October 2018. https://www.bellingcat.com/news /uk-and-europe/2018/10/02/anatoliy-chepiga-hero-russia-writing-wall/.

———. 2018. "Chemical Weapons and Absurdity: The Disinformation Campaign Against the White Helmets." Bellingcat, 18 December 2018. https://www.bellingcat.com/news/ mena/2018/12/18/chemical-weapons-and-absurdity-the-disinformation-campaign -against-the-white-helmets/.

———. 2018. "Full Report: Skripal Poisoning Suspect Dr. Alexander Mishkin, Hero of Russia." Bellingcat, 9 October 2018. https://www.bellingcat.com/news/uk-and-europe/2018 /10/09/full-report-skripal-poisoning-suspect-dr-alexander-mishkin-hero-russia/.

———. 2019. "The GRU Globetrotters: Mission London." Bellingcat, 28 June 2019. https:// www.bellingcat.com/news/uk-and-europe/2019/06/28/the-gru-globetrotters -mission-london/.

Benkler, Yochai, Robert Faris, and Hal Roberts. 2018. *Network Propaganda: Manipulation, Disinformation, and Radicalization in American Politics*. Oxford, UK: Oxford University Press.

Berger, Jonah. 2016. *Contagious: Why Things Catch On*. New York: Simon and Schuster.

Boo, Katherine. 2007. "Difficult Journalism That's Slap-Up Fun." In *Telling True Stories: A Nonfiction Writers' Guide from the Nieman Foundation at Harvard University*, edited by Mark Kramer and Wendy Call. New York: Plume.

Borneman, J. 1992. *Belonging in the Two Berlins: Kin, State, Nation*. Cambridge, UK: Cambridge University Press.

Bosch, T. 2017. "Twitter Activism and Youth in South Africa: The Case of #Rhodes-MustFall." *Information, Communication & Society* 20, no. 2: 221–32.

Bosman, Johan. 2013. "Mali." In *The Indigenous World 2013*. Copenhagen: International Work Group for Indigenous Affairs.

Bourgeot, André. 1995. *Les societies touarègues: Nomadisme, identité, résistances*. Paris: Éditions Karthala.

Brin, Sergei, and Lawrence Page. 1998. "The Anatomy of a Large-Scale Hypertextual Web Search Engine." *Computer Networks and ISDN Systems* 30: 107–17. Also available at http://infolab.stanford.edu/pub/papers/google.pdf.

Brodkin, Jon. 2011. "The MIME Guys: How Two Internet Gurus Changed E-mail Forever." *Networked World*, 1 February 2011. https://www.networkworld.com/article/2199390/ uc-voip/the-mime-guys—how-two-internet-gurus-changed-e-mail-forever.html.

Browning, Christopher. 2018. "The Suffocation of Democracy." *New York Review of Books*, 25 October 2018. https://www.nybooks.com/articles/2018/10/25/suffocation -of-democracy/.

Carrier, P. 2005. *Holocaust Monuments and National Memory Cultures in France and*

Germany since 1989: The Origins and Political Function of the Vél' d'Hiv' in Paris and the Holocaust Monument in Berlin. New York and Oxford: Berghahn.

Castells, Manuel. 1997. *The Information Age: Economy, Society and Culture*, vol. 2, *The Power of Identity*. Malden, MA: Blackwell.

Chakravarti, Sonali. 2008. "More than 'Cheap Sentimentality': Victim Testimony at Nuremberg, the Eichmann Trial and Truth Commissions." *Constellations* 15, no. 2: 223–35.

Champion, Matthew. 2015. *Medieval Graffiti: The Lost Voices of England's Churches*. London: Ebury.

Chen, Yongxi, and Anne Cheung. 2017. "The Transparent Self Under Big Data Profiling: Privacy and Chinese Legislation on the Social Credit System." *Journal of Comparative Law* 12, no.2: 356–78.

Cleaver, Harry. 1995. "The Zapatistas and the Electronic Fabric of Struggle." Unpublished ms. https://la.utexas.edu/users/hcleaver/zaps.html.

———. 1998. "The Zapatista Effect: The Internet and the Rise of an Alternative Political Fabric." *Journal of International Affairs* 51, no. 2: 621–40.

———. 1998. "The Zapatistas and the Electronic Fabric of Struggle." Unpublished ms. http://www.schoolsforchiapas.org/wp-content/uploads/2014/07/HarryCleaver.pdf.

Cohen, Jared, and Jacquelline Fuller. 2013. "Fighting Human Trafficking." Google. org (blog), 9 April 2013. https://www.blog.google/outreach-initiatives/google-org /fighting-human-trafficking/.

Coleman, Gabriella. 2012. *Coding Freedom: The Ethics and Aesthetics of Hacking*. Princeton, NJ: Princeton University Press.

Collier, George, and Elizabeth Lowery Quaratiello. 2005. *Basta! Land and the Zapatista Rebellion in Chiapas*. Oakland, CA: Food First.

Comaroff, Jean, and John Comaroff, eds. 2006. *Law and Disorder in the Postcolony*. Chicago, IL: University of Chicago Press.

Conger, Kate. 2018. "Ex-Google Employee Urges Lawmakers to Take On Company." *New York Times*, 26 September 2018. https://www.nytimes.com/2018/09/26/technology /google-privacy-china-congress.html.

Costanza-Chock, Sasha. 2014. *Out of the Shadows, Into the Streets! Transmedia Organizing and the Immigrant Rights Movement*. Cambridge, MA: MIT Press.

Couture, Stéphane. 2019. "The Ambiguous Boundaries of Computer Source Code and Some of Its Political Consequences." In *digitalSTS: A Field Guide for Science & Technology Studies*, edited by Janet Vertesi and David Ribes. Princeton, NJ: Princeton University Press.

Cozzolino, Francesca. 2017. *Peindre pour agir: Muralisme et politique en Sardaigne*. Paris: Karthala.

Crampton, A. 2001. "The Voortrekker Monument, the Birth of Apartheid, and Beyond." *Political Geography* 20: 221–46.

Crapanzano, V. 2011. *The Harkis: The Wound That Never Heals*. Chicago: University of Chicago Press.

Crete-Nishihata, Masashi, Jakub Dalek, Etienne Maynier, and John Scott-Railton. 2018.

Spying on a Budget: Inside a Phishing Operation with Targets in the Tibetan Community. Online report, part 3. The Citizen Lab. https://citizenlab.ca/2018/01/spying-on -a-budget-inside-a-phishing-operation-with-targets-in-the-tibetan-community /#part3.

Crettiez, Xavier, and Pierre Piazza. 2014. *Murs rebelles: Iconographie nationaliste contestaire: Corse, Pays Basque, Irlande du Nord*. Paris: Karthala.

Crimp, Douglas. 1990. *AIDS Demo Graphics*. Seattle: Bay Press.

DaPonte, Jason. 2015. "Taking Back the Web." *Index on Censorship* 44, no. 3: 86–88. http://journals.sagepub.com/doi/10.1177/0306422015605732.

Davis, Sarah. In press. " 'Don't Complain of the Dark': The Persistence of Chinese Rights Defenders." In *The Oxford Handbook of Law and Anthropology*, edited by Mark Goodale, Marie-Claire Foblets, Olaf Zenker, and Maria Sapignoli. Oxford, UK: Oxford University Press.

de Azevedo Cunha, Mario Viola, Norberto Nuno Gomes de Andrade, Lucas Lixinski and Lúcio Tomé Féteira eds. 2013. *New Technologies and Human Rights: Challenges to Regulation*. Surrey, UK: Ashgate.

de Gaay Fortman, Bastian. 2011. *Political Economy of Human Rights: The Quest for Relevance and Realization*. New York: Routledge.

della Porta, Donatella, ed. 2007. "The Global Justice Movement: An Introduction." In *The Global Justice Movement: Cross-National and Transnational Perspectives*, edited by Donatella della Porta, 1–28. Oxford and New York: Paradigm.

Der Spiegel online. 2005. "Remembering the Holocaust; Extracting Meaning from Concrete Blocks." 2 May 2005. http://www.spiegel.de/international/spiegel/remem bering-the-holocaust-extracting-meaning-from-concrete-blocks-a-354837.html.

Deutsche Welle. 2015. "Bundesregierung nennt Namibia-Massaker 'Völkermord.' " 10 July 2015. http://www.dw.com/de/bundesregierung-nennt-namibia-massaker -völkermord/a-18576575.

Die Welt. 2002. "Hört endlich auf mit eurer weißen Arroganz." Interview with Sam Nujoma. 2 December 2002. https://www.welt.de/print-welt/article282118/Hoert-endlich-auf-mit-eurer weissen-Arroganz.html.

Doshi-Velez. Finale, and Mason Kortz. 2017. *Accountability of AI under the Law: The Role of Explanation*. Working paper. Berkman Klein Center Working Group on Explanation and the Law, Berkman Klein Center for Internet & Society. http://nrs .harvard.edu/urn-3:HUL.InstRepos:34372584.

Dubberley, Sam, Alexa Koenig, and Daragh Murray, eds. 2020. *Digital Witness: Using Open Source Information for Human Rights Investigation, Documentation and Accountability*. Oxford, UK: Oxford University Press.

du Pisani, A. 2010. "The Discursive Limits of SWAPO's Dominant Discourses on Anti-Colonial Nationalism in Postcolonial Namibia—A First Exploration." In *The Long Aftermath of War—Reconciliation and Transition in Namibia*, edited by A. du Pisani, R. Kössler and W. A. Lindeke, 1–40. Freiburg: Arnold Bergstraesser Institut.

The Economist. 2018. "How 'Identitarian' Politics Is Changing Europe: White, Right and

Pretentious." 31 March 2018. https://www.economist.com/news/europe/21739668
-having-learned-identity-politics-left-right-wingers-are-shaping.

The Economist. 2006. "This Land Is My Land: Yet Another Land-Claim Dispute Turns
Ugly and Sheds a Spotlight on the Failure of Canada's Policies toward Its Aboriginal
People." 14 September 2006. http://www.economist.com/node/7911293.

Eicker, S. 2009. *Der Deutsch-Herero-Krieg und das Völkerrecht: Die Völkerrechtliche Haf-
tung der Bundesrepublik Deutschland für das Vorgehen des Deutschen Reiches gegen
die Herero in Deutsch-Südwestafrika im Jahre 1904 und ihre Durchsetzung vor einem
nationalen Gericht.* Frankfurt: Peter Lang.

Elias, Norbert. 2001. *The Society of Individuals.* Edited by M. Schröter and translated by
E. Jefcott. New York and London: Continuum.

Erichsen, Caspar. 2008. *"What the Elders Used to Say": Namibian Perspectives on the Last
Decade of German Colonial Rule.* Windhoek: Meinert.

Eubanks, Virginia. 2017. *Automating Inequality: How High-Tech Tools Profile, Police,
and Punish the Poor.* New York: St. Martin's.

Eysenbach, G., J. Powell, M. Englesakis, C. Rizo, and A. Stern. 2004. "Health-Related
Virtual Communities and Electronic Support Groups: Systematic Review of the
Effects of Online Peer-to-Peer Interactions." *British Medical Journal* 328 (15 May):
1166–70.

Falk, Richard. 2001. Preface. In *The Right to Self-Determination: Collected Papers and
Proceedings of the First International Conference on the Right to Self-Determination
and the United Nations, Geneva 2000,* edited by Y. N. Kly and D. Kly. Atlanta, GA:
Clarity.

Farkas, Yvette, ed. 2012. *Toronto Graffiti: The Human behind the Wall.* Toronto, ON:
Toronto Graffiti. www.torontograffiti.ca.

Federman, Sarah. 2018. "The 'Ideal Perpetrator': The French National Railways and the
Social Construction of Accountability." *Security Dialogue* 49, no. 5: 327–44.

Fischer, F., and N. Čupić. 2015. *Die Kontinuität des Genozids: Die Europäische Modern und der Völkermord an den Herero und Nama in Deutsch-Südwestafrika.* Berlin:
AphorismA.

Ford, Martin. 2018. *Architects of Intelligence: The Truth about AI from the People Building It.* Birmingham, UK: Packt.

Förster, L. 2008. "From 'General Field Marshall' to 'Miss Genocide': The Reworking of
Traumatic Experiences among Herero-Speaking Namibians." *Journal of Material
Culture* 13, no. 2: 175–94.

Förster, S., S. Frank, G. Krajewsky, and J. Schwerer. 2016. "Negotiating German Colonial
Heritage in Berlin's Afrikanisches Viertel." *International Journal of Heritage Studies*
22 (7): 515–29.

Freedland, Jonathan, 2018. "Russia's Brazen Lies Mock the World. How Best to Fight
for the Truth?" *The Guardian,* 15 September 2018. https://www.theguardian.com
/commentisfree/2018/sep/15/lies-russia-rt-salisbury-suspects-putin.

Free Fahad (blog). 2007. "Free Fahad: The Untold Story of Syed Fahad Hashmi." 7 June
2007. https://freefahad.wordpress.com/2007/06/07/hello-world/.

Froehling, Oliver. 1997. "The Cyberspace 'War of Ink and Internet' in Chiapas, Mexico." *Geographical Review* 87, no. 2: 291–307.

G7 2017. 2017. "G7 Declaration on Responsible States Behavior in Cyberspace." Lucca, Italy, 11 April. https://www.mofa.go.jp/files/000246367.pdf.

Gal, Olivier. 2016. *Truskool: Une histoire du graffiti à Toulouse*. Biarritz: Éditions Atlantica.

Gallagher, Ryan. 2018. "Google China Prototype Links Searches to Phone Numbers." *Intercept*, 14 September 2018. https://theintercept.com/2018/09/14/google-china -prototype-links-searches-to-phone-numbers/.

Garrido, Maria, and Alexander Halavais. 2003. "Mapping Networks of Support for the Zapatista Movement: Applying Social-Networks Analysis to Study Contemporary Social Movements." In *Cyberactivism: Online Activism in Theory and Practice*, edited by Martha McCaughey and Michael Ayers. New York and London: Routledge.

Gellner, David, and Eric Hirsch, eds. 2001. *Inside Organizations: Anthropologists at Work*. Oxford and New York: Berg.

Gensburger, Sarah, and Sandrine Lefranc. 2017. *À quoi servent les politiques de mémoire?* Paris: Les Presses Sciences Po.

Giedion, Sigfried. (1948) 2013. *Mechanization Takes Command: A Contribution to Anonymous History*. Minneapolis: University of Minnesota Press.

Goldsmiths' Center for Research Architecture and SITU Studio, 2010, "Report: Summary of Findings on the April 17, 2009 Death of Bassem Ibrahim Abu Rahma, Bil'in," http://www.situstudio.com/blog/wp-content/uploads/2010/07/Abu _Rahma_report_web.pdf.

Goody, Jack, ed. 1975. *Literacy in Traditional Societies*. Cambridge, UK: CambridgeUniversity Press.

———. 1977. *The Domestication of the Savage Mind*. Cambridge, UK: Cambridge University Press.

———. 1986. *The Logic of Writing and the Organization of Society*. Cambridge, UK: Cambridge University Press.

———. 1987. *The Interface between the Written and the Oral*. Cambridge, UK: Cambridge University Press.

Goody, Jack, and Ian Watt. 1968. "The Consequences of Literacy." In *Literacy in Traditional Societies*, edited by Jack Goody. Cambridge: Cambridge University Press.

Graham, B., G.J. Ashworth, and J.E. Tunbridge. 2000. *A Geography of Heritage: Power, Culture and Economy*. London: Hodder Arnold.

Gramsci, Antonio. (1929–1932) 1975. *Quaderni del carcere*, edited by Valentino Gerrantana. 4 vols. Torino: Einaudi.

Greenberg, Andy. 2016. "Google's New YouTube Analysis App Crowdsources War Reporting." *Wired*, 20 April 2016, 10:17 a.m. https://www.wired.com/2016/04/googles -youtube-montage-crowdsources-war-reporting/.

Gross, Daniel. 2018. "The Troubling Origins of the Skeletons in a New York Museum: Thousands of Herero People Died in a Genocide. Why Are Herero Skulls in the American Museum of Natural History?" *New Yorker*, 24 January 2018. https://

www.newyorker.com/culture/culture-desk/the-troubling-origins-of-the-skeletons
-in-a-new-york-museum.

Günther, Klaus. 2008. "Legal Pluralism or Dominant Concept of Law: Globalization as
a Problem of Legal Theory." *No Foundations* 5: 5–21.

Habermas, Jürgen. 1988. "Concerning the Public Use of History." Translated by J. Lea-
man. *New German Critique* 44: 40–50.

———. 1989 [1962]. *The Structural Transformation of the Public Sphere: An Inquiry into
a Category of Bourgeois Society*. Translated by Thomas Burger and Frederick Lawrence. Cambridge, MA: MIT Press.

Hägler, M. 2006. "Manche kämpfen weiter gegen die Hereros." *Taz*, 25 October 2006.
http://www.taz.de/!360590/.

Halbfinger, David. 2018. "A Day, a Life: When a Medic Was Killed in Gaza, Was It an
Accident?" *New York Times*, 30 December 2018. https://www.nytimes.com/2018
/12/30/world/middleeast/gaza-medic-israel-shooting.html.

Harari, Yuval Noah. 2017. *Homo Deus: A Brief History of Tomorrow*. New York: Harper
Collins.

Hatton, Celia. 2014. "China's Internet Vigilantes and the 'Human Flesh Search Engine.'"
BBC News, 28 January 2014. https://www.bbc.com/news/magazine-25913472.

Hawad. 2017. *Furigraphie: Poésies 1985–2015*. Translated by Hélène Claudot-Hawad.
Paris: Gallimard.

Herman, Edward, and Noam Chomsky. 2002. *Manufacturing Consent: The Political
Economy of the Mass Media*. New York: Pantheon.

Higgins, Eliot. 2017. *Op zoek naar die waarheid: Finding Truth in a Post-Truth World*.
Amsterdam and Antwerp: *Uitgeverij* Business Contact.

Hindman, M., K. Tsioutsiouliklis, and J. Johnson. 2003. "Googlearchy: How a Few
Heavily-Linked Sites Dominate Politics on the Web." Paper presented at the annual
meeting of the American Political Science Association, Philadelphia, PA.

Hochschild, Adam. 2012. *King Leopold's Ghost: A Story of Greed, Terror and Heroism in
Colonial Africa*. London: Pan.

Horton, Helena. 2016. "Microsoft Deletes 'Teen Girl' AI after It Became a Hitler-Loving
Sex Robot within 24 Hours." *The Telegraph*, 24 March 2016. http://www.telegraph.
co.uk/technology/2016/03/24/microsofts-teen-girl-ai-turns-into-a-hitler-loving
-sex-robot-wit/.

Hsu, Wendy. 2014. "Digital Ethnography toward Augmented Empiricism: A New Methodological Framework." *Journal of Digital Humanities* 3, no. 1, 43–64.

Hull, I. 2005. *Absolute Destruction: Military Culture and the Practices of War in Imperial
Germany*. Ithaca, NY: Cornell University Press.

Humphrey, John P. 1983. "The Memoirs of John P. Humphrey." *Human Rights Quarterly*
5: 387–439.

ICC (International Criminal Court). 2017. Situation in Libya in the Case of *The
Prosecutor v. Mahmoud Mustafa Busayf Al-Werfalli*. Warrant of Arrest. ICC-01/11–
01/17. 15 August 2017. https://www.icccpi.int/CourtRecords/CR2017_05031
.PDF.

———. 2018. Situation in Libya in the Case of *The Prosecutor v. Mahmoud Mustafa Busayf Al-Werfalli*. Second Warrant of Arrest. ICC-01/11–01/17, 4 July 2018. https://www.icc -cpi.int/CourtRecords/CR2018_03552.PDF.

Indigenous4earth. 2006. "La convention de Pau—25 juin 2006." africultures (website). http://www.africultures.com/php/index.php?nav=murmure&no=2985.

Innis, Harold. (1951) 2008. *The Bias of Communication*. 2nd ed. Toronto: University of Toronto Press.

Irle, J. 1906. *Die Herero: Ein Beitrag zur Landes-, Volks- und Missionskunde*. Gütersloh: C. Bertelsmann.

Isin, Engin, and Evelyn Ruppert. 2015. *Being Digital Citizens*. London and New York: Rowman & Littlefield.

ITU (International Telecommunications Union). 2017. *Measuring the Information Society Report*, vol.1. Geneva: International Telecommunications Union.

———. 2018. *The State of Broadband 2018: Broadband Catalyzing Sustainable Development*. https://www.itu.int/dms_pub/itu-s/opb/pol/S-POL-BROADBAND.19-2018-PDF-E.pdf.

James, David, and Rachel MacKenzie. 2018. "Stalking and Harrassment." In *The Routledge International Handbook of Human Aggression*, edited by Jane Ireland, Philip Birch, and Carol Ireland, 172–82. New York: Routledge.

Jessup, Philip C. 1950. *Transnational Law*. New Haven, CT: Yale University Press.

Johns, Fleur. 2016. "Global Governance through the Pairing of List and Algorithm." *Environment and Planning D: Society and Space* 34, no. 1: 126–49.

———. 2017. "Data Mining as Global Governance." In *The Oxford Handbook of Law, Regulation and Technology*, edited by Roger Brownsword, Eloise Scotford, and Karen Yeung. Oxford Handbooks Online. DOI: 10.1093/oxfordhb/9780199680832.013.56. Oxford, UK: Oxford University Press.

Johnson, N. 1995. "Cast in Stone: Monuments, Geography, and Nationalism." *Environment and Planning D: Society and Space* 13: 51–65.

Kaba, Lasine. 1974. *The Wahhabiyya: Islamic Reform and Politics in French West Africa*. Evanston, IL: Northwestern University Press.

Kang, Cecilia. 2018. "Executives Pull Out of Saudi Conference." *New York Times*, 12 October 2018. https://www.nytimes.com/2018/10/19/technology/tech-executives-saudi-conference.html.

Karpf, David. 2016. *Analytic Activism: Digital Listening and the New Political Strategy*. Oxford, UK: Oxford University Press.

Keenan, Jeremy. 2004. *The Lesser Gods of the Sahara: Social Change and Contested Terrain amongst the Tuareg of Algeria*. London, UK, and Portland, OR: Frank Cass.

———. 2010. "Resisting Imperialism: Tuareg Threaten US, Chinese and other Foreign Interests." In *Tuareg Society Within a Globalized World: Saharan Life in Transition*, edited by Anja Fischer and Ines Kohl. London: I. B. Tauris. 209–30.

———. 2012. "How Washington Helped Foster the Islamist Uprising in Mali." *New Internationalist*, 1 December 2012. https://newint.org/features/2012/12/01/us-terrorism -sahara.

Kesselring, Rita, Elif Babül, Mark Goodale, Tobias Kelly, Ronald Niezen, Maria Sapignoli, and Richard A. Wilson. 2017. "The Future of the Anthropology of Law. Emergent Conversation." *PoLAR: Political and Legal Anthropology Review Online*, 10 February 2017. https://polarjournal.org/2017/02/10/emergent-conversations-part-6/.

Kiberd, Roisin. 2016. "The Nightmarish Online World of 'Gang-Stalking': The Internet Creates an Echo chamber for Delusions and Conspiracies." *Vice*, 22 July 2016. https://www.vice.com/en_us/article/aeknya/the-nightmarish-online-world-of-gang-stalking.

Kingsbury, Benedict. 2000. "Reconstructing Self-Determination: A Relational Approach." In *Operationalizing the Right of Indigenous Peoples to Self-Determination*, edited by Pekka Aikio and Martin Scheinin. Turku, Finland: Institute for Human Rights, Abo Akademi University.

Kisting, D. 2017. "Namibia's Wealth Redistribution Plan May Benefit Elite Minority." *Bloomberg*, 29 August, https://www.bloomberg.com/news/articles/2017-08-28/namibia -black-ownership-plan-risks-repeating-south-africa-errors.

Kitchin, Rob. 2017. "Thinking Critically about Researching Algorithms." *Information, Communication & Society* 20, no. 1: 14–29.

Klarsfeld, Beate, and Serge Klarsfeld. 2015. *Mémoires*. Paris: Fayard/Flammarion.

———. 2018. *Hunting the Truth: Memoirs of Beate and Serge Klarsfeld*. New York: Farrar, Straus, and Giroux.

Kleinman, Arthur, and Joan Kleinman. 1997. "The Appeal of Experience; The Dismay of Images: Cultural Appropriations of Suffering in Our Times." In *Social Suffering*, edited by Arthur Kleinman, Vena Das, and Margaret Lock. Berkeley: University of California Press.

Klute, Georg. 2010. "Kleinkrieg in der Wuste: Nomadische Kriegsfuhrung und die 'Kultur des Krieges' bei den Tuareg." In *Die Komplexitat der Kriege: Globale Gesellschaft und Internationale Beziehungen*, edited by Thomas Jaeger, 188–222. Wiesbaden: VS Verlag/Springer.

Kochhar, Rakesh. 2015. "Seven in Ten People Globally Live on $10 or Less per Day." Pew Research Center, 23 September 2015. http://www.pewresearch.org/fact -tank/2015/09/23/seven-in-ten-people-globally-live-on-10–or-less-per-day/.

Kocka, J. 1990. "Deutsche Identität und historischer Vergleich: nach dem 'Historikerstreit.'" *Almanach: ein Lesebuch*, vol. 3, 65–84, Bonn: Jackwerth & Welker.

Kohl, Ines, and Anja Fischer. 2010 "Tuareg Moving Global: An Introduction." In *Tuareg Society Within a Globalized World: Saharan Life in Transition*, edited by Anja Fischer and Ines Kohl, 1–8. London: I. B. Tauris.

Korster, M. 2013. "Mau Mau Reparations, Memorialization and Kenya's Future: Reflections after Fifty Years of Independence." 2013 Conference Proceedings, Kenyan Scholar and Studies Association. http://kessa.org/conference_proceedings /2013_conference_proceedings.

Kössler, R. 2007. "Facing a Fragmented Past: Memory, Culture and Politics in Namibia." *Journal of Southern African Studies* 33, no. 2: 361–82.

———. 2015. *Namibia and Germany: Negotiating the Past*. Windhoek: University of Namibia Press.

Kozinets, Robert. 2010. *Netography: Doing Ethnographic Research Online*. London: Sage.

Kristof, Nicholas. 2018. "Be Outraged by America's Role in Yemen's Misery." *New York Times*, 26 September 2018. https://www.nytimes.com/2018/09/26/opinion/yemen-united -states-united-nations.html.

Krug, Nora. 2018. *Heimat: Ein deutsches Familienalbum*. Munich: Penguin.

Kubal, T. 2008. *Cultural Movements and Collective Memory: Christopher Columbus and the Rewriting of the National Origin Myth*. New York: Palgrave Macmillan.

Kuss, S. 2004. "Der Herero-Deutsche Krieg und das deutsche Militär: Kriegsursachen und Kriegsverlauf." In *Namibia–Deutschland: eine geteilte Geschichte. Widerstand, Gewalt, Erinnerung*, edited by L. Förster, D. Henrichsen, and M. Bollig, 62–77, Munich: Minerva.

Lackmann, Thomas. 2015. "Warum heißt die Mohrenstrasse Mohrenstrasse?" *Der Tagesspiegel*, 24 January 2015. http://www.tagesspiegel.de/berlin/streit-um-strassen- namen-warum-heisst-die-mohrenstrasse-mohrenstrasse/7332938.html.

Lamberti, Elena. 2012. *Marshall McLuhan's Mosaic: Probing the Literary Origins of Media Studies*. Toronto, ON: University of Toronto Press.

Land, Molly. 2016. "Democratizing Human Rights Fact-Finding." In *The Transformation of Human Rights Fact-Finding*, edited by Philip Alston and Sarah Knuckey. Oxford, UK: Oxford University Press.

Land, Molly, and Jay Aronson, eds. 2018. *New Technologies for Human Rights Law and Practice*. Cambridge, UK: Cambridge University Press.

Lang, Mabel. 1988. *Graffiti in the Athenian Agora*. Rev. ed. Athens: American School of Classical Studies at Athens.

La Stampa. 2015. "Umberto Eco: 'Con i social parola a legioni di imbecilli.'" 11 June 2015. https://www.lastampa.it/2015/06/10/cultura/eco-con-i-parola-a-legioni-di- imbecilli-XJrvezBN4XOoyooh98EfiJ/pagina.html.

Latour, Bruno. 2005. *Reassembling the Social: An Introduction to Actor-Network Theory*. Oxford, UK: Oxford University Press.

Lecocq, Baz, and Georg Klute. 2013. "Tuareg Separatism in Mali." *International Journal* 8, no. 3: 424–34.

Lee, Seung. 2016. "Google Targets 'The Fog of War' in Syria." *Newsweek*, 20 April 2016. https://www.newsweek.com/google-targets-fog-war-syria-450408.

Lei, Ya-Wen. 2018. *The Contentious Public Sphere: Law, Media, and Authoritarian Rule in China*. Princeton, NJ: Princeton University Press.

Lippmann, Walter. 1922. *Public Opinion*. New York: Harcourt, Brace.

Lipstadt, Deborah. 2011. *The Eichmann Trial*. New York: Schocken.

Lu, Catherine. 2017. *Justice and Reconciliation in World Politics*. Cambridge, UK: Cambridge University Press.

Lunacek, Sarah. 2010. "Ambiguous Meanings of *Ikufar* and the Role in Development Projects." In *Tuareg Society Within a Globalized World: Saharan Life in Transition*, edited by Anja Fischer and Ines Kohn, 191–208. London and New York: I. B. Tauris.

Lusignan, Guy de. 1969. *French-Speaking Africa since Independence*. New York: Frederick A. Praeger.

Macdonald, Nancy. 2003. *The Graffiti Subculture: Youth, Masculinity, and Identity in London and New York*. London: Palgrave MacMillan.

Macdonald, S. 2009. *Difficult Heritage: Negotiating the Nazi Past in Nuremberg and Beyond*. London and New York: Routledge.

MacKinnon, Rebecca. 2012. "The Netizen." *Development* 55, no. 2: 201–4.

Macrakis, Kristie. 2008. *Seduced by Secrets: Inside the Stasi's Spy-Tech World*. Cambridge, UK: Cambridge University Press.

Manovich, Lev. 2001. *The Language of New Media*. Cambridge, MA: MIT Press.

Mantovani, Marco. 2017. *Vecchia Scuola: Graffiti Writing a Milano*. Milan: Stradedarts.

Margulies, Peter. 2016. "Surveillance by Algorithm: The NSA, Computerized Intelligence Collection, and Human Rights." *Florida Law Review* 68, no. 4: 1045–117.

Marrus, Michael. 2002. *The Unwanted: European Refugees from the First World War Through the Cold War*. Philadelphia: Temple University Press.

McIntyre, Lee. 2018. *Post-Truth*. Cambridge, MA: MIT Press.

McLaughlin, Timothy. 2018. "Facebook Blocks Accounts of Facebook's Top General, Other Military Leaders." *Washington Post*, 27 August 2018. https://www.washingtonpost.com/world/asia_pacific/facebook-blocks-accounts-of-myanmars-top-general-other-military-leaders/2018/08/27/da1ff440-a9f6-11e8-9a7d-cd30504ff902_story.html.

McLuhan, Marshall. (1962) 2011. *The Gutenberg Galaxy*. Toronto, ON: University of Toronto Press.

———. (1964) 2017. *Understanding Media: The Extensions of Man*. Edited by W. Terrence Gordon. Berkeley, CA: Gingko Press.

McPherson, Ella. 2018. "Risk and the Pluralism of Digital Human Rights Fact-Finding and Advocacy." In *New Technologies for Human Rights Law and Practice*, edited by Molly Land and Jay Aronson, 188–214. Cambridge, UK: Cambridge University Press.

Melber, H. 2007. "Poverty, Politics, Power and Privilege: Namibia's Black Economic Elite Formation." In *Transitions in Namibia: Which Changes for Whom?*, edited by H. Melber, 110–29, Uppsala: Nordic Africa Institute.

Melber, H., and R. Kössler. 2015. "Wer B sagt, muss auch A sagen: Völkermord an Herero und Nama: Ein Debattenbeitrag von Mit-Initiatoren der Kampagne "Völkermord ist Völkermord." *Internationale Politik und Gesellschaft*, 17 July 2015. http://www.ipg-journal.de/rubriken/aussen-und-sicherheitspolitik/artikel/wer-b-sagt-muss-auch-a-sagen-1001/.

Mériadec, Raffray. 2013. *Touaregs: La révolte des hommes bleues 1857–2013*. Paris: Economica.

Merry, Sally Engle. 2006. "Anthropology and International Law." *Annual Review of Anthropology* 35: 99–116.

———. 2006. *Human Rights and Gender Violence: Translating International Law into Local Justice*. Chicago: University of Chicago Press.

———. 2016. *The Seductions of Quantification: Measuring Human Rights, Gender Violence, and Sex Trafficking*. Chicago: University of Chicago Press.

———. In press. "Law in Practice: The Problem of Compliance and the Turn to Quantification." In *The Oxford Handbook of Law and Anthropology*, edited by Mark Goodale, Marie-Claire Foblets, Olaf Zenker, and Maria Sapignoli. Oxford, UK: Oxford University Press.

Metz, Cade. 2012. "How the Queen of England Beat Everyone to the Internet." *Wired*, 25 December. https://www.wired.com/2012/12/queen-and-the-internet/.

Microsoft. n.d. *A Digital Geneva Convention to Protect Cyberspace*. Microsoft Policy Papers. https://query.prod.cms.rt.microsoft.com/cms/api/am/binary/RW67QH.

Miller, Elissa. 2018. "Egypt Leads the Pack in Internet Censorship Across the Middle East." *MENASource*, 28 August 2018. Atlantic Council. https://www.atlanticcouncil.org/blogs/menasource/egypt-leads-the-pack-in-internet-censorship-across-the-middle-east.

Milnor, Kristina. 2014. *Graffiti and the Literary Landscape in Roman Pompeii*. Oxford, UK: Oxford University Press.

Morgan, Andy. 2012. "The Causes of the Uprising in Northern Mali." Think Africa Press, 6 February 2012. http://thinkafricapress.com/mali/causes-uprising-northern-mali-tuareg.

Mortensen, Mette. 2011. "When Citizen Photojournalism Sets the News Agenda: Neda Agha Soltan as a Web 2.0 Icon of Post-election Unrest in Iran." *Global Media and Communication* 7, no. 1: 4–16.

MoveOn. 2018. "Students Pledge to Refrain from Interviewing with Google until Commitment not to Pursue Future Tech Military Contracts (e.g. Project Maven)." https://petitions.moveon.org/sign/students-pledge-to-refrain.

Moyn, Samuel. 2012. *The Last Utopia: Human Rights in History*. Cambridge, MA: Harvard University Press.

———. 2018. *Human Rights in an Unequal World*. Cambridge, MA: Harvard University Press.

Murphy, Thérèse, ed. 2009. *New Technologies and Human Rights*. Oxford, UK: Oxford University Press.

Murthy, Dhiraj. 2008. "Digital Ethnography: An Examination of the Use of New Technologies for Social Research." *Sociology* 42, no. 5: 837–55.

The National. 2013. "Tuareg Insurgents in Mali Raped Hundreds of Women and Girls, UN Says." Bloomberg News. 31 January 2013. https://www.thenational.ae/world/africa/tuareg-insurgents-in-mali-raped-hundreds-of-women-and-girls-un-says-1.34852.

Netzpolitik.com (website). 2015. "Über uns." https://netzpolitik.org/about-this-blog/.

Niezen, Ronald. 1990. "The Community of Helpers of the Sunna: Islamic Reform among the Songhay of Gao (Mali)." *Africa* 60, no. 3: 399–424.

———. 1991. "Hot Literacy in Cold Societies: A Comparative Study of the Sacred Value of Writing." *Comparative Studies in Society and History* 33, no. 2: 225–54.

———. 2003. *The Origins of Indigenism: Human Rights and the Politics of Identity*. Berkeley: University of California Press.

———. 2005. "Digital Identity: The Construction of Virtual Selfhood in the Indigenous Peoples' Movement." *Comparative Studies in Society and History* 45, no. 2: 532–51.

———. 2010. *Public Justice and the Anthropology of Law*. Cambridge, UK: Cambridge University Press.

———. 2013. Internet Suicide: Communities of Affirmation and the Lethality of Communication." *Transcultural Psychiatry* 50, no. 2: 303–22.

————. 2014. "Gabriel Tarde's Publics." *History of the Human Sciences* 27, no. 2: 41–59.

————. 2017. "The Future of the Anthropology of Law: PoLAR's Sixth Emergent Conversation." *PoLAR*, 10 February 2017. https://polarjournal.org/ronald-niezen -mcgill-university/.

————. 2017. "Il volto pubblico dell'ingiustizia. Attivismo indigeno e insurrezioni dei Tuareg in Mali" (The public face of injustice: Indigenous activism and Tuareg insurrection in Mali). In *La questione indigena in Africa* (The indigenous question in Africa), edited by Maria Sapignoli, Robert Hitchcock, and Gaetano Mangiamele. Milan: UNICOPLI.

————. 2017. "Speaking for the Dead: The Memorial Politics of Genocide in Namibia and Germany." *International Journal of Heritage Studies* 24, no. 5 (22 December). https://doi.org/10.1080/13527258.2017.1413681.

————. 2017. *Truth and Indignation: Canada's Truth and Reconciliation Commission on Indian Residential Schools.* 2nd ed. Toronto: University of Toronto Press.

Noble, Safiya Umoja. 2018. *Algorithms of Oppression: How Search Engines Reinforce Racism.* New York: New York University Press.

Nordbruch, C. 2006. *Völkermord an den Herero in Deurtsch-Südwestafrika? Widerlegung einer Lüge.* Tübingen: Grabert.

Obk, Comer. 2017. *Marqué à vie! 30 ans de graffiti "vandal."* Paris: Éditions Da Real.

Obst, Patricia, and Jana Stafurik. 2010. "Online We are All Able Bodied: Online Psychological Sense of Community and Social Support Found through Membership in Disability-Specific Websites Promotes Well-being for People Living with a Physical Disability." *Journal of Community & Applied Social Psychology* 20, no. 6: 525–31.

Olusoga, D., and C. Erichsen. 2011. *The Kaiser's Holocaust: Germany's ForgottenGenocide.* London: Faber and Faber.

Ó Muíneacháin, Conn. 2012. "Thanks, Al Gore." Podcast #30. Technology.ie. http://tech nology.ie/thanks-al-gore-podcast-30/.

O'Neil, Cathy. 2017. *Weapons of Math Destruction: How Big Data Increases Inequality and Threatens Democracy.* New York: Broadway Books.

Osman, Mohamed. 2018. "ICC Suspect Al-Werfalli 'Escapes' from Prison in Libya." *International Justice Monitor*, August 2018. https://www.ijmonitor.org/2018/08/icc -suspect-al-werfalli-escapes-from-prison-in-libya/.

Ott, Brian. 2017. "The Age of Twitter: Donald J. Trump and the Politics of Debasement." *Critical Studies in Media Communication* 34, no. 1: 59–68.

Pandolfi, Paul. 2001. "Les touaregs et nous: Une relation triangulaire?" *Ethnologies Comparées* no. 2 (printemps). http://www.vitaminedz.com/articlesfiche/0/91.pdf.

Papacharissi, Zizi. 2015. *Affective Publics: Sentiment, Technology, and Politics.* Oxford, UK: Oxford University Press.

Pariser, Eli. 2011. *The Filter Bubble: How the New Personalized Web Is Changing What We Read and How We Think.* New York: Penguin.

Pearl, Judea. 2019. "The Limitations of Opaque Learning Machines." In *Possible Minds: 25 Ways of Looking at AI*, edited by John Brockman. New York: Penguin.

Peraldi-Mittelette, Paul. 2018. "Touaregs 2.0." *Netcom: Networks and Communication Studies* 32, no. 3/4: 331–46. https://journals.openedition.org/netcom/3037.

Phillips, Whitney. 2015. *This Is Why We Can't Have Nice Things: Mapping the Relationship between Online Trolling and Mainstream Culture.* Cambridge, MA: MIT Press.

Pils, Eva, Taisu Zhang, Isabel Hilton, Xiao Qiang, Edward Friedman, Isaac Stone Fish, Jerome Cohen, Robert Daly, Thomas Kellogg, and Hu Yong. 2016. "'Rule by Fear?': A ChinaFile Conversation." ChinaFile Conversation, 18 February 2016. ChinaFile. http://www.chinafile.com/conversation/rule-fear.

Pinter, Harold. 2005. Nobel Lecture. NobelPrize.org. Nobel Media AB 2019. https://www.nobelprize.org/prizes/literature/2005/pinter/25621-harold-pinter-nobel-lecture-2005/.

Postman, Neil. 1985. *Amusing Ourselves to Death: Pubic Discourse in the Age of Show Business.* New York: Penguin.

———. 1992. *Technopoly: The Surrender of Culture to Technology.* New York: Vintage.

Pritchard, V. 1967. *English Medieval Graffiti.* Cambridge, UK: Cambridge University Press.

Rasmussen, Susan. 2006. *Spirit Possession and Personhood among the Kel Ewey Tuareg.* Cambridge, UK: Cambridge University Press.

Raso, Filippo, Hannah Gilligoss, Vivek Krishnamurthy, Christopher Bavitz, and Levin Kim. 2018. *Artificial Intelligence & Human Rights: Opportunities & Risks.* Cambridge, MA: Berkman Klein Center for Internet & Society, Harvard University.

Resta, Giorgio, ed. 2010. "Il problema dei processi mediatici nella prospettiva del dirittocomparato." In *Il rapporto tra giustizia e mass media: quali regole per quali soggetti,* edited by Giorgio Resta. Naples: Editoriale Scientifica.

Riles, Annelise. 2001. *The Network Inside Out.* Ann Arbor: University of Michigan Press.

———, ed. 2006. *Documents: Artifacts of Modern Knowledge.* Ann Arbor: University of Michigan Press.

Roberts, Julian, and Loretta Stalans. 2000. *Public Opinion, Crime, and Criminal Justice.* Boulder, CO: Westview.

Roberts, Siobhan. 2018. "The Yoda of Silicon Valley: Donald Knuth, Master of Algorithms, Reflects on 50 Years of His Opus-in-Progress, 'The Art of Computer Programming.'" *New York Times,* 17 December 2018. https://www.nytimes.com/2018/12/17/science/donald-knuth-computers-algorithms-programming.html.

Rodríguez, Clemencia. 2011. *Citizens' Media against Armed Conflict: Disrupting Violence in Columbia.* Minneapolis, MN: University of Minnesota Press.

Ronfeldt, David, and John Arquilla. 2001 "Emergence and Influence of the Zapatista Social Netwar." In *Networks and Netwars: The Future of Terror, Crime, and Militancy,* edited by John Arquilla and David Ronfeldt, 171–99. Santa Monica, CA: RAND.

Rottenburg, Richard. 2009. *Far-Fetched Facts: A Parable of Development Aid.* Translated by Allison Brown and Tom Lampert. Cambridge, MA: MIT Press.

Rovira, Guiomar. 2009. *Zapatistas sin fronteras: Las redes de solidaridad con Chiapas y el altermundismo.* Tlalpan, Mexico: Ediciones Era.

———. 2017. *Activismo en red y multitudes conectadas: Communicación y acción en la era de Internet.* Barcelona: Icaria Editorial; Antrazyt.

RT News, 2018. "'Suck My Balls': NATO-Funded Blogger Eliot Higgins Checkmates Critics with Watertight Argument." 14 June 2018. https://www.rt.com/news/429609 -eliot-higgins-suck-my-balls/.

Ruser, Nathan. 2018. *How to Scrape Interactive Geospatial Data.* 5 September 2018. https:// www.bellingcat.com/resources/how-tos/2018/09/05/scrape-interactive-geospatial -data/.

Samuels, David. 2019. "Is Big Tech Merging With Big Brother? Kinda Looks Like It." *Wired*, 23 January 2019. https://www.wired.com/story/is-big-tech-merging-with-big -brother-kinda-looks-like-it/.

Sapignoli, Maria. 2017. "A Kaleidoscopic Institutional Form: Expertise and Transformation in the UN Permanent Forum on Indigenous Issues." In *Palaces of Hope: The Anthropology of Global Organizations*, edited by Ronald Niezen and Maria Sapignoli. Cambridge UK: Cambridge University Press.

———. 2018. *Hunting Justice: Displacement, Law, and Activism in the Kalahari.* Cambridge, UK: Cambridge University Press.

Sapignoli, Maria, and Ronald Niezen. 2020. "Global Legal Institutions." In *Oxford Handbook of Anthropology and Law*, edited by Mark Goodale, Marie-Claire Foblets, Maria Sapignoli, and Olaf Zenker. Oxford, UK: Oxford University Press.

Sarkin, Jeremy, and Carly Fowler. 2008. "Reparations for Historical Human Rights Violations: The International and Historical Dimensions of the Alien Torts Claims Act Genocide Case of the Herero of Namibia." *Human Rights Review* 9: 331–60.

Sarogni, Emilia. 2004. *La Donna Italiana: 1861–2000, il lungo cammino verso i dritti.* Milan: Gruppo editoriale il Saggiatore.

Savage, K. 2009. *Monument Wars: Washington D.C., the National Mall and the Transformation of the Memorial Landscape.* Berkeley: University of California Press.

Schildkrout, Enid. 1995. "Museums and Nationalism in Namibia." *Museum Anthropology* 19, no. 2: 65–77.

Schneier, Bruce. 2018. *Click Here to Kill Everybody: Security and Survival in a Hyper-Connected World.* New York and London: W. W. Norton.

Schreiber, Wolfgang. 2013. "Nichtstaatliche bewaffnete lokale und internationale Gruppen in Mali." In *Wegweiser zur Geschichte Mali*, edited by Martin Hofbauer and Philipp Münch, Zentrum für Militärgeschichte und Soziawissenschaften der Bundeswehr. Paderborn: Ferdinand Schöningh.

Sejnowski, Terrence. 2018. *The Deep Learning Revolution.* Cambridge, MA: MIT Press.

Shandler, Jeffrey. 1997. *While America Watches: Televising the Holocaust.* New York: Oxford University Press.

Shane, Scott, and Daisuke Wakabayashi. 2018. "'The Business of War': Google Employees Protest Work for the Pentagon." *New York Times*, 4 April 2018. https://www. nytimes.com/2018/04/04/technology/google-letter-ceo-pentagon-project.html.

Silva, L., and P. Mota Santos. 2012. "Ethnographies of Heritage and Power." *International Journal of Heritage Studies* 18, no. 5: 437–43.

Simonite, Tom. 2016. "Moore's Law Is Dead. Now What?" *MIT Technology Review*, 13 May 2016. https://www.technologyreview.com/s/601441/moores-law-is-dead-now -what/.

Small, Deborah, George Loewenstein, and Paul Slovic. 2007. "Sympathy and Callousness: The Impact of Deliberative Thought on Donations to Identifiable and Statistical Victims." *Organizational Behavior and Human Decision Processes* 102, no. 2: 143–53.

Smith, Laurajane. 2006. *Uses of Heritage*. London and New York: Routledge.

Smith, M. G. 1974. *Corporations and Society: The Social Anthropology of Collective Action*. New Brunswick, NJ, and London: Aldine Transaction.

Soares, Benjamin. 2005. *Islam and the Prayer Economy: History and Authority in a Malian Town*. Ann Arbor: University of Michigan Press.

———. 2013. "Islam in Mali since the 2012 Coup." *Cultural Anthropology Online*, 10 June 2013, http://production.culanth.org/fieldsights/321-islam-in-mali-sincethe -2012-coup.

Statt, Nick. 2016. "Ellen Pao Launches Advocacy Group to Improve Diversity in the Tech Industry." *Verge*, 3 May 2016. https://www.theverge.com/2016/5/3/11579798/ellen-pao -project-include-announced-tech-diversity-initiative.

Stein, Mary Beth. 2008. "Stasi with a Human Face? Ambiguity in 'Das Leben der Anderen.'" *German Studies Review* 31, no. 3: 567–79.

Steinhauser, G. 2017. "Germany Confronts the Forgotten Story of Its Other Genocide." *Wall Street Journal*, 28 July 2-17. https://www.wsj.com/articles/germany-confronts -the-forgotten-story-of-its-other-genocide-1501255028.

Strick, Benjamin. 2018. *How to Identify Burnt Villages by Satellite Imagery—Case Studies from California, Nigeria, and Myanmar*. Bellingcat. https://www.bellingcat.com/resources /how-tos/2018/09/04/identify-burnt-villages-satellite-imagery%E2%80%8A -case-studies|-california-nigeria-myanmar/.

Süddeutsche Zeitung. 2015. "Ein-Man-Nachrichtenagentur." 1 June 2015. https://archive .is/S690b.

Sunstein, Cass. 2009. *Going to Extremes: How Like Minds Unite and Divide*. Oxford: Oxford University Press.

———. 2017. *#Republic: Divided Democracy in the Age of Social Media*. Princeton, NJ: Princeton University Press.

Tarde, Gabriel. 1893. *Les transformations du droit: Étude sociologique* [The Transformations of Law: A Sociological Study]. Paris: Félix Alcan.

———. 1902. *Psychologie économique*. Vol. 2. Paris: Félix Alcan.

———. 2011. *On Communication and Social Influence: Selected Papers*. Translated and edited by Terry Clark. Chicago: University of Chicago Press.

Tate, Winnifred. 2007. *Counting the Dead: The Culture and Politics of Human Rights Activism in Columbia*. Berkeley: University of California Press.

———. 2013. "Proxy Citizenship and Transnational Advocacy: Columbian Activists from Putumayo to Washington, DC," *American Ethnologist* 40, no. 1: 55–70.

The Telegraph. 2017. "Chinese Tourists Arrested for Giving Hitler Salute outside Reichstag

Building in Berlin." 6 August 2017. http://www.telegraph.co.uk/news/2017/08/06
/chinese-tourists-arrested-giving-hitler-salute-outside-reichstag/.

Toler, Aric. 2018. *Creating an Android Open Source Research Device on Your PC*. Bellingcat. https://www.bellingcat.com/resources/how-tos/2018/08/23/creating-android
-open-source-research-device-pc/.

Trouillot, Michel-Rolf. 1995. *Silencing the Past: Power and the Production of History*.
Boston: Beacon.

Tufekci, Zeynep. 2017. *Twitter and Tear Gas: The Power and Fragility of Networked Protest*. New Haven, CT: Yale University Press.

———. 2018. "It's the (Democracy-Poisoning) Golden Age of Free Speech." *Wired*, 16
January 2018. https://www.wired.com/story/free-speech-issue-tech-turmoil-new
-censorship/.

UNDESA (United Nations Department of Economic and Social Affairs). 2018. "Frontier Technologies for a Sustainable Future." *Voice* 22, no. 10 (October). https://
www.un.org/development/desa/undesavoice/wp-content/uploads/sites/49/2018/10
/DESAVoiceOctober2018.pdf.

UNDGACM (United Nations Department for General Assembly and Conference Management). n.d. "Information and Communications Technology Section." http://
www.un.org/depts/DGACM/funcicts.shtml.

UNHCR (United Nations Human Rights Council). 2018. *Report of the Independent International Fact-Finding Mission on Myanmar*. UN doc. no. A/HCR/39/64. https://
www.ohchr.org/_layouts/15/WopiFrame.aspx?sourcedoc=/Documents/HRBodies
/HRCouncil/FFM-Myanmar/A_HRC_39_64.docx.

United Nations. 2017. "Dag Hammarskjöld Library: Update on UN Digitalization Pro-
gramme." 2 May. https://library.un.org/content/update-un-digitization-programme.

———. 2018. "The Secretary-General's Strategy on New Technologies." https://www
.un.org/en/newtechnologies/images/pdf/SGs-Strategy-on-New-Technologies.pdf.

UNSCD (United Nations Strategic Communications Division). n.d. "Strategic
Communications Division." http://www.un.org/en/sections/department-public-in-
formation/department-public-information/strategic-communications/index.html.

Vaidhyanathan, Siva. 2018. *Anti-Social Media: How Facebook Disconnects Us and Under-
mines Democracy*. Oxford, UK: Oxford University Press.

Vansina, J. 2010. *Being Colonized: The Kuba Experience in Rural Congo, 1880–1960*.
Madison, WI: University of Wisconsin Press

Viejo-Rose, D. 2011. *Reconstructing Spain: Cultural Heritage and Memory after Civil
War*. Brighton, UK: Sussex Academic Press.

Viggiani, E. 2014. *Talking Stones: The Politics of Memorialization in Post-Conflict Northern Ireland*. New York and Oxford: Berghahn.

Volkan, Vamik. 2001. "Transgenerational Transmission and Chosen Traumas: An Aspect of Large-Group Identity." *Group Analysis* 34, no. 1: 79–97.

Wallace, Marion. 2011. *A History of Namibia: From the Beginning to 1990*. London:
Hurst.

Wang, Lucy Lu, Gabriel Stanovsky, Luca Weihs, and Oren Etzioni. 2019. *Gender Trends in Computer Science Authorship*. Preprint. Seattle, WA: Allen Institute for Artificial Intelligence. https://arxiv.org/abs/1906.07883.

Weber, Max. 1946. *From Max Weber: Essays in Sociology*. Edited by Hans Gerth and C. Wright Mills. Oxford, UK: Oxford University Press.

Weisbord, Noah. 2019. *The Crime of Aggression: The Quest for Justice in an Age of Drones, Cyberattacks, Insurgents, and Autocrats*. Princeton, NJ: Princeton University Press.

Weizman, Eyal. 2014. "Introduction: Forensis." In *Forensis: The Architecture of Public Truth*. London: Sternberg Press and Forensic Architecture.

Werner, Wolfgang. 1993. "A Brief History of Land Dispossession in Namibia." *Journal of Southern African Studies* 19, no. 1: 135–46.

Williams, Patricia. 2019. "Why Everyone Should Care About Mass E-carceration." *Nation*, 13 May 2019. https://www.thenation.com/article/surveillance-prison -race-technology/.

Wimsatt, William. 2001. *Bomb the Suburbs: Graffiti, Race, Freight-Hopping and the Search for Hip-Hop's Moral Center*. Berkeley, CA: Soft Skull Press.

Woodside, Arch, Suresh Sood, and Kenneth Miller. 2008. "When Consumers and Brands Talk: Storytelling Theory and Research in Psychology and Marketing." *Psychology and Marketing* 25, no. 2: 97–145.

Wylie, Christopher. 2019. *Mindf*ck: Cambridge Analytica and the Plot to Break America*. New York: Random House.

Young, J. 1992. "The Counter-Monument: Memory against Itself in Germany Today." *Critical Inquiry* 18, no. 2: 267–96.

Young, Robert. 2001. *Postcolonialism: A Historical Introduction*. Oxford, UK: Blackwell.

Zayadin, Hiba. 2018. "We Have Two Months to Save Israa al-Ghomgham's Life." Human Rights Watch, 28 August 2018. https://www.hrw.org/news/2018/08/28/ we-have-two-months-save-israa-al-ghomghams-life.

Zill, R. 2011. "'A True Witness of Transience': Berlin's Kaiser-Wilhelm-Gedächtni-skirche and the Symbolic Use of Architectural Fragments in Modernity." *European Review of History/Revue européenne d'histoire* 18, no. 5–6: 811–27.

Zimmerman, Kim Ann, and Jesse Emspak. 2017. "Internet History Timeline: ARPANET to the World Wide Web." Live Science. https://www.livescience.com/20727-internet -history.html.

Zuboff, Shoshana. 2019. *The Age of Surveillance Capitalism: The Fight for a Human Future at the New Frontier of Power*. New York: Public Affairs.

Zuern, E. 2012. "Memorial Politics: Challenging the Dominant Party's Narrative in Namibia." *Journal of Modern African Studies* 50, no. 3: 493–518.

———. 2017. "Namibia's Monuments to Genocide." *Dissent*, 13 June 2017. https://www .dissentmagazine.org/blog/namibia-genocide-monuments-reparations-germany.

Zumbansen, Peer. 2008. "Transnational Law." *Comparative Research in Law & Political Economy*. Research Paper no. 9/2008.

Zweig, Stefan. 1998. *Sternstunden der Menschheit*. Frankfurt: Fischer.

Index

Page numbers in italics refer to figures.

Aboubacrine, Saoudata, 153
Aboubakrine, Mariam, 146, 154
Abu-Rahma, Bassem, 99
Acharatoumane, 148–49
ACT UP (AIDS Coalition to Unleash Power), 50–51
Afrotak TV cyberNomads, 158
Ag Fadil, Boubacar, 148–49
Ag Ghali, Iyad, 140, 141
Agisoft Metashape (photogrammetric processor), 105
AI (artificial intelligence): deep learning and algorithms, 67–70, 78–79; and ethics, 73; Microsoft Twitter experiment, 60
Albachir, Aboubacar, 152
algorithms, 68–70, 78–79, 89, 112, 116–20
Alliance for Democracy and Tolerance against Extremism and Violence (BFDT), 158
Al-Qaida au Maghreb Islamique (AQMI), 136, 147, 148, 151–52
Alternative für Deutschland (German political party), 158, 167
Amazon, 68
Amnesty International, 31, 90, 99
analog activism, 45–51, 189
Anonymous, 17, 92
Ansar ad-Dīn, 140–41, 148, 151
Apple, 68
Arab Spring, 4, 56, 72
Arendt, Hannah, 156, 164
artificial intelligence (AI), deep learning and algorithms, 67–70, 78–79; and ethics, 73; Microsoft Twitter experiment, 60

Asal, Victor, 117
al-Assad, Bashar, 93
Atlantic Council's Digital Forensic Research Lab (DFRLab), 98
Austin, Regina, 123
authoritarianism, 8, 36, 82–88, 119, 157. See also state control
authoritarian liberalism, 18
Awake: A Dream from Standing Rock (documentary film), 129

Baryin, Gael, 146
British Broadcasting Service (BBC), 87, 94
Bellingcat, 91–107, 196; challenges and possibilities, 106–7; collaborations, 97–98; establishment and early investigations, 93–95; goal, 95–96; method application, 101–3, 105–6; open-source tools, overview, 91–93; workshops, 95, 96–97, 101
Berger, Jonah, 111
Berlin Philharmonic, 126
Bloom, Orlando, 126
Boo, Katherine, 111
Bornstein, Nathaniel, 52
Brave Bull Allard, LaDonna, 127
break boundaries, 44
Brin, Sergei, 117
British colonialism, 163, 168
Browning, Christopher, 119
Bureau of Investigative Journalism, 99

Castells, Manuel, 56
CBS Evening News, 51

The authorized representative in the EU for product safety and compliance is:
Mare Nostrum Group
B.V Doelen 72
4831 GR Breda
The Netherlands

www.ingramcontent.com/pod-product-compliance
Lightning Source LLC
Chambersburg PA
CBHW020844270326

41928CB00006B/535

9781503612631